THE EARTH BENEATH THE SEA

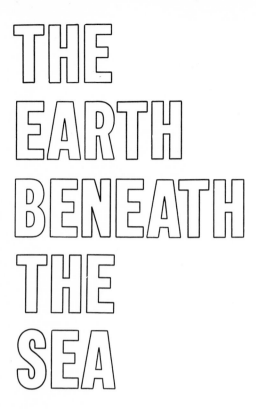

THE EARTH BENEATH THE SEA

Revised Edition

by Francis P. Shepard

The Johns Hopkins Press

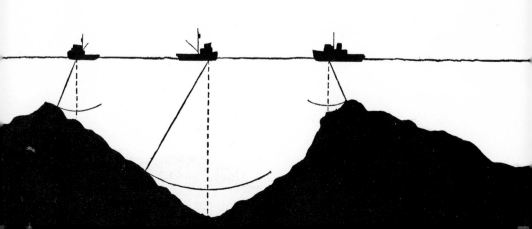

Originally published, 1959
Second printing, 1960
Third printing, 1961

Revised edition, 1967
Second printing, 1968

to Elizabeth

PREFACE

This book was written partly in response to a desire expressed by many friends to have a book that they could read without a background of technical knowledge. It is essentially a popularization of my book *Submarine Geology*, avoiding discussion of complicated and technical aspects.

Another purpose in writing the book was to show the fun as well as the tribulations in trying to develop a new field of science. When I started out forty years ago, I thought of myself as being virtually the only geologist studying the ocean floor. I have learned since then that there had previously been various other attempts, some of them long before I got started. Just recently I ran across a book by a French geologist, M. Deleese, entitled *Lithologie du Fond des Mers (The Rocks of the Sea Bottom)*, which was published in 1866. It covered many of the same fields as my early work, but seems to have got lost, like so many other pioneer attempts. A German, K. Andree, also published a book, *Geologie des Meeresbodens (Geology of the Sea Bottom)*, in 1920. Finally, a Russian, M. V. Klenova, brought out a book on the same subject in the same year as mine, and she has informed me that it appeared before mine. Someone has always planted the flag ahead of you.

Being almost alone in a field does have some difficulties. My training was certainly inadequate. We would look with horror on a prospective student who tried to enroll in Scripps Institution of Oceanography for graduate studies with a background like mine. I have had to obtain most of my education since I received my doctor's degree. I have been fortunate, however, in having been surrounded for the past twenty years by specialists in all branches of the study of the sea, so that I have learned from them and, to considerable extent, with them. In writing this book I have drawn freely in various chapters both from their pioneering work and from their advice.

Among those who have been particularly helpful to me are R. S. Arthur, J. R. Curray, R. L. Fisher, D. L. Inman, H. W. Menard, F. B Phleger, R. W. Raitt, and G. A. Rusnak of Scripps Institution of Oceanography; R. S. Dietz of ESSA; E. L. Hamilton of the Navy Electronics Laboratory; H. S. Ladd of the United States Geological Survey; and Blair Kinsman and D. W. Pritchard of Chesapeake Bay Institute. Considerable help has been given me in my attempt to write this book for popular consumption by the late Mrs. Ruth Young Manar and by my wife. The drafting has been almost entirely the result of the enthusiastic work by J. R. Moriarty. For the faithful copying and recopying of the chapters my appreciation and sympathy are expressed to Mrs. Betty Sanborn. Many of the results reported in this book have come from work carried on under contracts with the Office of Naval Research and the American Petroleum Institute and National Science Foundation.

The second edition has attempted to update the earlier treatment. An enormous amount of progress has been made in the past eight years, so it seemed advisable to acquaint readers with the principal developments during that period. A few sections have been rewritten and the rest revised, but without changing the main content of the book. Some of the speculative ideas in the earlier edition have been removed because of new information, and other new ideas have been introduced. However, we have not yet reached a state in the study of marine geology that would give it the status of a firmly established science. Much is yet to be accomplished. Margaret Miller and J. R. Curray have been very helpful in making this revision.

September, 1966

CONTENTS

THE EARTH BENEATH THE SEA

INTRODUCTION

The study of the geology of the ocean floor, commonly referred to as submarine or marine geology, has as its purpose the explanation of the continental shelves, the slopes, the canyons, and the deep-sea floor (Fig. 1). This study involves a total area of approximately 72 per cent of the earth's surface. The nature and origin of the sediments that only partially cover these vast submerged areas are also involved in the study, as are the rocks that underlie this sediment. The attack of geologists on this territory of the sea has necessarily

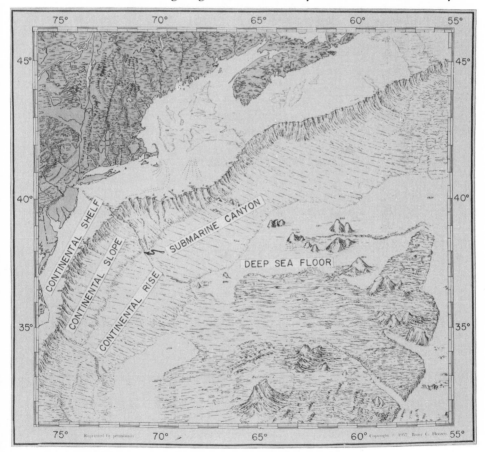

Figure 1. *Physiographic diagram showing the continental shelf, continental slope, continental rise, and various types of deep-sea floor off the eastern seaboard. Vertical exaggeration approximately times 20. Courtesy of Bruce Heezen of Lamont Geological Observatory and Marie Tharp of Lamont Geological Observatory.*

differed from that on land, since the area studied has been mostly out of the range of vision and too deep to allow the application of the Brunton compass and the hammer, so important in land studies. Instead, dredging, coring, and random photography by means of cameras in protected cylinders have had to act as substitutes. It is only in recent years that a few geologists have started exploring the shallow margins of the ocean by swimming with scuba, which consists of one or more cylinders of compressed air strapped to the back and a tube leading to the mouth for breathing. With these they have been able to apply direct methods to the zone where water depths are less than about 180 feet. The possibility of extending these direct studies to much greater depths came first from the development of the bathyscaph. This device, invented by Auguste Piccard in 1948, is literally an underwater balloon having a large envelope filled with gasoline to make it lighter than water for the ascent and, in the place of the gondola, a steel ball with thick windows and enough room inside for two observers. During the descent small steel shot are held in place by a magnet, but these can be released upon reaching the bottom to allow the balloon to rise. There are propellers and a motor, which are capable of giving the bathyscaph only very slow backward and forward motion for a short period.

Jacques Cousteau, in 1957, started another trend in deep-diving vehicles by developing his *Diving Saucer,* which consists primarily of two nine-foot saucer-shaped hulls welded together. The submarine is able to move independently into very narrow caves and canyons by means of water jets. It can explore the bottom to depths of 1,000 feet. The Navy's *Deep Jeep* can go down to 2,000 feet; the Westinghouse *Deep Star* has a practical exploration depth of 4,000 feet; Woods Hole's *Alvin,* capable of work at 6,000 feet, and Reynolds Submarine Services' *Aluminaut,* a larger vehicle with a diving range of 15,000 feet, are now all in operation and adding to information about the sea floor. An ambitious program aimed largely at sport fishermen is that of the Perry *Cubmobile;* most of these diving vehicles go only to 300-foot depths. Various other companies are developing deep-diving vehicles, and it can be anticipated that this field will have a great expansion in the near future.

The advance of man into the depths and of photography with automatic control to even greater depths is indicated in Figure 2. Despite these pioneer efforts, submarine geology is still largely dependent on observations made from vessels over an area where the scientist can judge the appearance of the bottom that he is exploring from echo-sounding profiles alone.

The difficulties of conducting shipboard operations that involve contact with the sea floor are little appreciated by those who have not been involved. The loss of expensive equipment is commonplace when one is grappling with rock ledges and pinnacles that cannot be seen and working from a rolling and pitching platform that is being carried along by currents of proportions difficult to evaluate. The great pressures of the depths cause apparatus to collapse, and the cold temperatures decrease the effect of lubrication so that parts "freeze"

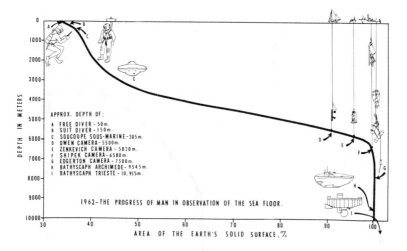

Figure 2. Hypsographic curve showing the area of the earth's solid surface above any given depth. Along the curve are the present means of directly viewing the sea floor. The percentage of the earth's surface that each method helped to make available for exploration can be determined along the bottom line. Courtesy of E. L. Hamilton, Navy Electronics Laboratory.

and the salt water and salt air corrode the metal unless very special care is maintained. Finally, many scientists are under a physical and mental strain caused by seasickness (now much less acute because of dramamine and bonamine) and by the dangers of accidents, that result from supporting tons of weight over the side of the vessel on rigging that is none too strong for the job. Thus, in practice seagoing geology lacks much of the glamor that has sometimes been attributed to it by landlubbers. It is hard, trying work, and results are still small.

Before World War II the seagoing geologists of the world could have had a convention in one office room. These were the men who were attempting to study the three-fourths of the earth's surface that is covered by the ocean. The war, along with all its horror, did at least stimulate many phases of science, of which marine, or submarine, geology was one. Having learned of the advantages to military activities of increased knowledge of the sea floor, the navies of the world have been generous in their contributions to the furtherance of this neglected branch of science. The activities of the Internationl Geophysical Year, and the International Indian Ocean expeditions have penetrated much farther into the dark abysses of the sea.

In some branches of science, such as nuclear physics, many of the great advances have been made by those investigators who have had time or taken time to think. The philosophic approach, however, has not been very successful in many phases of geology, because ideas have been based on too fragmentary a background of facts. Many of the problems of geology long debated in lecture halls have been solved by those who were willing to go out and dig up the contacts between rock formations. In the newest branch of geology, the study of the sea floor, there is still a woeful need for diggers.

Forty-five years ago I emerged from the cloistered halls of the University of Chicago with a suspicion that some of the hypotheses that I had been taught with almost religious fervor might not be as well founded as I had been led to suppose. At that time the geology textbooks dealt with the ocean mostly in the light of what geologists had learned from the world-encircling British *Challenger* Expedition of 1872-76 and from the study of marine sedimentary rocks obtained on land. The *Challenger* Expedition was certainly very productive, but because of the primitive means then existing for deep-sea soundings and for securing samples from the ocean bottom, the achievements of this one expedition were of course quite limited. Similarly the attempt to interpret conditions on the present sea floor from the study of ancient marine sedimentary rocks was obviously the wrong approach, for it was rather like trying to classify existing fauna and flora from a study of the fossil record without collecting any present-day forms.

The first indication I had that all was not the way it was described in the textbooks was in 1923 when I took a series of bottom samples in Massachusetts Bay, going out from the beach to where the water was a hundred feet or more in depth. I had been taught that sediment on the sea floor grew finer as one went out from the shore because of the decreasing power of the waves and currents. In almost all of my sample lines, however, the sediment, after some decrease in grain size, grew coarser near the outer end of the line. I began to look for evidence from other localities to see if this finding represented a rare exception to a seemingly well-established rule. I examined the notations concerning the ocean bottom on many navigation charts and again found many examples to correspond with my own findings. No one, apparently, seemed to have consulted the charts before establishing the generalizations. A visit to the National Museum of the Smithsonian Institution, in Washington, produced even more supporting evidence that the textbooks were wrong. Gathering dust in the attic of the National Museum were well-labeled bottles with thousands of samples obtained in the charting operations of the United States Coast and Geodetic Survey—samples that no one had taken the trouble to study. Here was proof that the shelf sediments do not grade outward from coarse to fine in crossing the continental shelf. In many places gravel was sampled from the outer shelf and mud from areas near the shore. More recently oceanographic institutions have provided numerous shelf samples. These also show the lack of outward gradation, but they have provided information that makes it possible to understand this curious enigma.

The taking of long cores in the sediments of the ocean floor has also led to revolutionary changes in geological ideas. Pipes up to 100 feet in length have been pushed into the bottom by using great weight on the coring device above the pipe, by utilizing hydrostatic pressure, and by using the principle of the piston to reduce friction. The cores taken from these pipes have shown that the deep-ocean floor has many layers indicative both of climatic changes and of the introduction of coarse sediments from shallow water by powerful currents that move down the continental slopes and out over the plains beyond. Study of the cores is providing a good yardstick for measuring the length of the various glacial stages of the great ice age.

The interest in drilling through the ocean floor was developed as a result of very successful petroleum operations on the continental shelf with drilling from platforms and, in part, from stationary barges. The *Mohole* project to drill through the earth's crust to the mantle, some three miles below the deep ocean floor, led to a considerable political hassel in Congress, and finally, in 1966, an abandonment of this ambitious project was made, at least for the time being. However, in its place has come drilling from barges through one to two thousand feet of the ocean floor, and this has already produced important results in what is known as the *Joides* drilling off the east coast of Florida. Plans for further Joides drilling in both major oceans are being made.

I remember having it impressed upon me in my university work that the ocean bottom was a flat monotonous surface with only a few rather minor irregularities around the margins. Here again was a philosophical concept— based, in this case, on the supposition that sediments falling on the sea floor had filled or smoothed the basins and buried most of the hills. The soundings of the *Challenger* Expedition, made at intervals of hundreds of miles, did little to confirm or deny this hypothesis. It was only in 1924, when echo soundings were first taken across an ocean basin on the German ship *Meteor,* that it was found that the irregularities of the ocean bottom might be as great as those of the continents. This has now been confirmed and amplified by countless fathograms made across the oceans of the world. Hundreds of high mountains have been discovered, especially in the Pacific Ocean, many of them forming parts of great mountain chains. Deep, elongated depressions (some of them known before the days of echo soundings) add to the major relief features that are now unfolding. On the other hand, new, extremely accurate echo-sounding devices now in use show that some of the flattest of all plains exist on the sea floor. The present idea of the origin of these plains is that they are due to deposition from sediment-laden currents that move rapidly down the slopes and out onto the deep-sea floor. This process would never have been suspected from the results of the *Challenger* Expedition.

Among the most perplexing and interesting features of the sea bottom are the submarine canyons. The extension of some of these deep valleys down the great marginal slopes of the continents has been known for over a century through the soundings made by the various marine surveyors, particularly those of the United States Coast and Geodetic Survey and the British Admiralty.

At first little attention was paid to these canyons, and most geologists who gave them any thought explained them as having resulted from local sinking of the land margins, bringing river valleys below the level of the ocean. In view of the great crustal movements that have occurred, elevating the sea bottom into high mountain ranges, this did not seem particularly puzzling. As one result of echo-sounding surveys, it has been discovered that the submarine canyons are very widespread, being found, in fact, along virtually all the coasts of the world. It has also been evident that some of the canyons have walls that are even deeper than those of the Grand Canyon of the Colorado. Because of these facts many geologists have become skeptical of the river origin of the sea canyons. Such an origin implies that the coasts of the world must have been depressed thousands of feet.

Quite recently Maurice Ewing and his group of scientists from Columbia University's Lamont Geological Observatory, located at Palisades, New York, began to trace some of these valleys, such as the submarine canyon off the Hudson River, out to the deep floor of the Atlantic. They found a troughlike valley 100 feet or more below its surroundings and extending out to where the ocean has depths of about 15,000 feet. Other shallow valleys have been found as seaward continuations of the submarine canyons of the Pacific, and some of them even extend down the oceanic deeps leading south from Baffin Bay and the Danish Strait, on either side of Greenland. These shallow valleys at great depths are not explained by anyone as having been cut by land rivers, and many geologists are convinced that submarine canyons in general are not submerged river valleys but are the result of some as-yet-unknown process taking place on the ocean bottom. But the fact remains that the well-surveyed inner canyons have a remarkable resemblance to river valleys.

Discoveries concerning the nature of sediments and rocks underlying the sea bottom have also brought great surprises to scientists in the last few years. Carefully computed estimates had been made of the thickness of sediments that mantle the deep-ocean floors. These estimates were based on the calculated amount of sediment now contributed annually to the ocean and on the estimates of the age of the oceans. These figures led the well-known Dutch marine geologist Philip Kuenen to compute in 1949 that there should be an average thickness of about two miles of sediment. However, when seagoing geophysicists started using the technique of the oil geologists to send sound waves through the ocean-floor sediment, it was found that the thickness was far less than supposed. In fact, in some places there was virtually no sediment. Elsewhere, a few thousand feet of sediment were all that the records indicated. Several explanations of this enigma have now been made, but there is no assurance that any of these are correct. Deep drilling will undoubtedly provide the answers.

When the investigations were carried still deeper into the crust below the sediment cover, a different type of rock was found under the oceans from that which exists under the continents. The speed of sound in the rocks underlying the oceans is much faster than in the rocks under the continents at the same crustal depth. As a result of these geophysical investigations, we are making some progress in understanding the reason that the floor of the ocean lies at a much lower level than that of the continents. The rocks with the higher speed of sound are probably heavier, and hence the crust under the oceans has sunk in relation to that under the continents.

Even deep drilling has been used as a means of investigating the history of the ocean. One of the long-controverted subjects in geology has been the origin of the ringlike atoll coral reefs so common in the South Pacific. A hundred years ago Charles Darwin suggested that these atolls represented former volcanic islands that had submerged slowly and upon which coral animals had developed their colonies, growing up to the surface at a rate about equal to that of the sinking of the islands. This was long a hotly debated hypothesis, and it is only in recent years that the drilling conducted by the United States Navy and the United States Geological Survey in preparation for the hydrogen bomb tests at Eniwetok helped solve the problem, showing that Darwin's ideas were fundamentally correct.

Waves and currents play an important role in modifying the sea floor, much more than was formerly supposed. Until very recent years ocean currents were thought to die out or become negligible at depths of a few hundred feet. We have found, however, that among the core samples brought up from the deep-sea floor there are clean sand and even gravel layers, both indicative of strong currents at great depths. Currents with velocities comparable at least to those of land rivers are now being considered as developing from time to time on the oceanic marginal slopes and continuing down to the deep-ocean floor. Waves, also supposed in the past to be limited to shallow water in their effect on the bottom, are now thought to exist on surfaces several thousand feet below sea level, although they are of different character than surface waves. Ripple marks of the type caused by waves are currently being found in flashlight photographs taken on the summits of submarine mountains at these great depths.

The serious erosion problems that have come from the building of jetties and other artificial coastal structures have led to a combined engineering and geological investigation of the marine processes that operate along the shore lines. In recent years there has come to light much new important information, which may lead to the more intelligent development of harbors along such inhospitable coasts as California and Florida, where natural harbors are at a premium. The extensive photographing of the deep ocean floors by Lamont Geological Observatory and other oceanographic institutions has recently led to the discovery that deep currents move along the slopes as well as down them, and in many places these are capable of transporting sediments and leaving ripple-marked surfaces.

In some of our early studies of shore processes at Scripps Institution we helped dispel the old myth about the dangers of undertow. Although our measurements show some slight net outward movements along the bottom, the really dangerous currents, which cause so many drownings, move seaward at the surface to an even greater degree than they do along the bottom.

Thus the branch of science that deals with the geology of the sea floor and of the oceanic margins is developing slowly but surely. As we get longer cores penetrating deeper into the bottom sediments, make more deep-water current measurements, take more ocean-floor photographs, and actually see more of the ocean floor with improved, more maneuverable deep submergence vehicles, the picture will undoubtedly change considerably. Right now we have a fairly good grasp of the problem.

In summary, we have learned during the last few decades that the deep-ocean floor has tremendous mountains and valleys. The margins have canyons as deep as the deepest on land. The sediments are of an amazing complexity even in deep water, where they were supposed to be monotonously similar. Currents and waves of a special type are operating at the greatest depths. The sedimentary rocks on land are being matched with recent sediments on the sea floor, making it possible to distinguish the environments in which these rocks were deposited. Gradually the ocean bottom is yielding its wealth to man with great promise for the future when land reserves of petroleum, cobalt, nickel, manganese, and no doubt other products become scarce.

CHAPTER I

WAVES AND CURRENTS MODIFY THE SEA FLOOR

Almost every square inch of the ocean floor, from the shore line to the greatest deeps, comes under the influence of currents or of some type of wave motion. The ordinary waves, which are created by wind blowing over the surface of the water, are restricted in their sphere of activity to depths of less than a few hundred feet. The combined effects of the various types of waves and currents are of fundamental importance in understanding the extraordinary nature of the sea floor, which all of the new methods of exploration are revealing to us. For that reason these processes will be considered first to give a background for a discussion of the underwater provinces.

Significance of Waves

Because of their importance to mankind, waves induced by the wind blowing on the water surface have been studied for more than a century by engineers, physicists, geologists, and marine architects. Accelerated advances came from the need of the various navies during World War II to know more about the action of waves, particularly at the shore. For one thing, landing operations were vitally influenced by wave conditions, as we learned to our sorrow when the Marines landed at Tarawa. It was often necessary to forecast the waves several days ahead of time in order to schedule a landing for a period when waves would give the least interference. The variation of wave height from place to place along the shores had to be understood in order to make landings where the smallest waves should be encountered. Detailed analyses were made of the wave origins and of the propagation of waves outside of storm centers. Landings were assisted by the forecasts, and the Allied operations in both the Atlantic and Pacific were clearly benefited by the studies. In the case of the Normandy invasion, D-Day, originally scheduled for June 5, 1944, was postponed one day because of the prediction of high waves. On succeeding days, almost hourly forecasts controlled the plans for landing of personnel and unloading of equipment. So close was the relation between wave conditions and permissible offloading that the wave predictions for a given day were essentially a quantitative estimate of that day's debarkation.

Since World War II, researchers in meteorology and oceanography have continued investigations to improve the early techniques and apply them to civilian needs. Forecasting is proving invaluable in designing and maintaining costly offshore structures, such as oil drilling platforms in the Gulf of Mexico and in the North Sea. The value lies not only in preventing damage to expensive structures and their auxiliary barges and tenders, but also in reducing the hazard to human lives by permitting evacuation of personnel prior to the arrival of storm waves and hurricanes.

Mechanics of Wave Motion

A detailed discussion of waves would be highly mathematical and beyond the scope of this book, but the nature of wave action may be clarified by a consideration of some of the properties of waves. A theoretical wave is illustrated in Figure 3, Section A, with the standard terminology. The high points are the *crests,* and the low point is the *trough.* The distance from crest to crest is the *wave length* and the vertical distance between crest and trough is the *wave height.* The time it takes two crests to pass a fixed point, for instance the end of a pier, is the *wave period.* The arrows in Figure 3 show the motion of water particles in the different parts of the wave. Thus at the wave crest the water is moving in the direction in which the wave is apparently advancing, but in the trough the water particles are moving in the opposite direction. The arrows to the left of the trough show that the water is rising to meet the

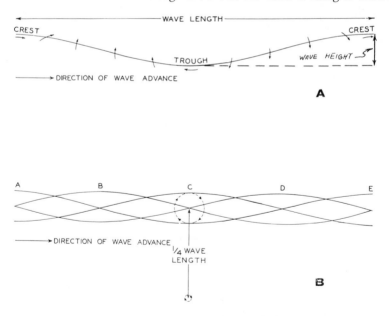

Figure 3. *Illustrating the nature of water motion in various parts of waves. Some of the common names applied to waves are included. Note that the water particle motion is much smaller at a depth equal to a quarter of a wave length.*

approaching crest. Similarly the water to the right of the trough has to fall in order to bring the trough to that position. The result of this motion is that in a wave period the water particles make a complete circle (see Fig. 3, Sec. B).

The preceding description refers to a theoretical wave. Actual waves are much more complicated. Under the action of the wind the sea humps up into very irregular bumps and hollows that run rapidly across the surface of the water, altering their shapes as they proceed. If one flies slowly near the surface of the ocean in a helicopter following an unusually high wave, one will see the wave die out in a short time and vanish among smaller waves. In a sea where the wind is causing waves to form, huge numbers of waves of different sizes are present, running in many directions. This is due to the great variability of the wind both in force and direction during a storm. The waves raised by the wind continue in approximately the same direction until they reach the land, which may be way beyond the limits of the storm. The waves in the storm area are called *sea,* and those that have outrun the storm are called *swell.* The swell is much more regular than the sea. This is because the longer-period waves run faster than the shorter ones, so that many of the small irregular waves are left behind nearer the storm area. The swell is far from regular, however, because of its source. If you watch individual swells come in along a pier, you will note an occasional series of large waves that are quite uniform in character and in period. These large swells are usually interspersed with a relatively large number of small swells. This is helpful in going through the breakers. After you have seen several large waves break, try to get through directly following the first small wave. The method is usually safe but not infallible.

The theoretical wave in Figure 3 shows that during each period water particles make a complete circuit, thus coming back to where they started. Actually the water has a net forward movement, very much like a car wheel that is spinning on a slippery surface and only moving forward a small fraction of the distance over which the wheel rotates. If you drop any buoyant object from the end of a pier or from a boat near shore, it will ordinarily move in toward the beach as the result of this net forward movement of the particles in the waves.

At depth the circular motion in a wave decreases rapidly (Fig. 3, Sec. B), becoming negligible at a depth equal to one half of the wave length. It is for this reason that submerged submarines are so little disturbed during storms and constitute excellent platforms for delicate scientific measurements. Wave motion can reach considerable depths under unusual storm conditions. At the mouth of the English Channel, swells have been known to move one-pound rocks into lobster pots at a depth of 180 feet. Off western Ireland, rocks weighing several hundred pounds have been moved in water 100 feet deep. Coarse sand is sometimes brought up from 150 feet of water and thrown against Bishop Rock Lighthouse on the English coast.

If you live along the southern California coast, you can often observe large swells coming into the shore despite the fact that there has been no sign of a storm in your vicinity. These swells are generally long period, eight to eighteen seconds. Some of the longer-period swells may have come from as far away as the South Pacific east of New Zealand. This *southern swell* is seen during Northern Hemisphere summers, because that is the period of the great winter storms in the Southern Hemisphere. The swell seen in southern California in winter generally comes from the northwest, often from storm centers just south of the Aleutian Islands. A European observer sees long swells coming in from various parts of the mid-Atlantic Ocean south of Greenland. The large booming waves that break at Casa Blanca on the west coast of Africa are an example of waves originating at a distant source, as experienced by many soldiers in their World War II landings in Morocco.

If you live along the eastern coast of the United States, you will rarely see the long rolling swell. Short, choppy waves are the common occurrence. Most of these are derived from local or nearby storm waves. The large swells more commonly hit the other side of the Atlantic because of the prevailing westerlies among the wave-generating winds.

Breakers. The general behavior of waves coming into shallow water can be approximated by some relatively simple physical laws. One effect is the transfer of the wave energy from a relatively deep column of water into one that is becoming shallower. The apparent wave velocity and the wage length decrease, while the wave height increases. When the velocity of the water at the crest exceeds the wave velocity, outrunning the rest of the wave, the wave breaks. The depth at which the wave breaks increases with the wave height and steepness. Thus short-period waves break in deeper water than long-period waves of the same breaker height,[1] as the short-period waves have greater steepness because their crests are closer together.

Inside the point where the waves break, the water moves largely onto the beach and back again, the *uprush* moving up as the wave comes in and the *backwash* returning. The violence of the backwash may knock a bather off his feet and carry him out into very turbulent and dangerous water, but there is very little, if any, undertow connected with this backwash.

The height of the breakers along the shore varies according to the submarine topography. The simplest case is illustrated in Figure 4. Here we see a shoreline that is straight, but on the adjacent sea floor there are ridges and submarine valleys. As the waves approach the shore and the water becomes shallow, the waves are slowed down except over the valleys, where the water is not shallow. The result is that the wave front becomes sinuous and the energy from an advancing wave moves from the valley onto the ridge on either

---

[1] Long-period waves increase their height before breaking, and therefore a five-foot long-period wave in deep water may produce a ten-foot breaker.

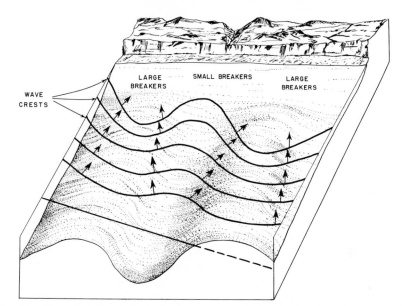

Figure 4. Illustrating the bend of the wave crest in relation to the submarine topography. Note that the energy of the wave crest moving up the submarine canyon is decreased due to divergence (spreading), whereas the energy on the ridges is increased due to convergence.

side, developing a nutcracker effect. Therefore, the breakers over the ridge are greatly augmented as they approach the shore, whereas over the valley, because of the loss of energy, the breakers become very small. A difference of ten times in breaker height is not unusual as the result of topography of this sort. In many places the waves approaching the shore over the valleys do not break at all except during very rough weather. Obviously this is a fine place to launch a boat. At La Jolla, California, the fishermen keep their skiffs inside such a valley. In western Europe the two fishing towns of Nazare, Portugal, and Cap Breton, France, are also located inside submarine valleys. Several California piers and jetties have been located to take advantage of these low-wave areas, for example, the jetties at Redondo and Hueneme and the pier at Moss Landing.[2]

The fall of the water in a breaker produces the wave's greatest erosional effect on the ocean floor. If you stand on a pier along the California coast and watch the breaking of a large long-period wave on a calm day, you can see that its crest curves over, leaving a hollow beneath, and descends with a free fall onto the smooth water surface below it. This *plunging breaker* (Fig. 5)

---

[2] These piers and jetties have had trouble caused by landslides into the submarine valleys (see Chap. VII).

Figure 5. Photograph of a plunging breaker with its hollow front, taken at La Jolla, California. This wave traveled for thousands of miles from a storm center before breaking on the California coast.

produces far more erosion where it breaks than does a *spilling breaker* (Fig. 6), which simply cascades down into a trough. The spilling breaker, however, stirs the bottom over a much wider area, so that it may have more total effect. If you swim along the bottom directly outside the area where plunging breakers actually break, you will find that as the wave crest passes overhead, the bottom is somewhat stirred, but relatively little material is thrown into suspension. Then, if you can stand the strain, come into the zone where the wave is breaking. Your vision will now be almost completely blocked by the violently roiled-up water. Measurements have indicated that the momentary velocities on the bottom in the breaker zone are several times those observed directly outside the breakers.

Figure 6. Photograph of a spilling breaker, taken at Marblehead Neck, Massachusetts, during a fifty-mile-an-hour blow.

Longshore Currents

Currents running with wave approach. The sediment stirred into suspension by the waves is transported by *longshore currents*. The cause of these currents is now well understood, although the currents themselves are still misinterpreted by many people who have been taught to believe in the largely fictitious undertow. It is in the slight net forward transport of water particles that we find the fundamental cause for the currents along the shore. When waves approach the shore diagonally, the water attempts to keep on moving in the direction of this approach. Since the water movement is stopped by the land, the water particles are deflected along the shore. Thus, if the waves are coming in from the left, the motion is ordinarily to the right, and vice versa. The resulting flow is called a longshore current, and it may develop a velocity of as much as one or more miles per hour, often exceeding the speed potential of an average swimmer.

Currents running opposite to wave approach. In some cases currents actually move in a direction opposite to that from which the waves are approaching. This results from the convergence that pushes waves together, building up the surface over a ridge (Figs. 4, 7). The result of this convergence is a current

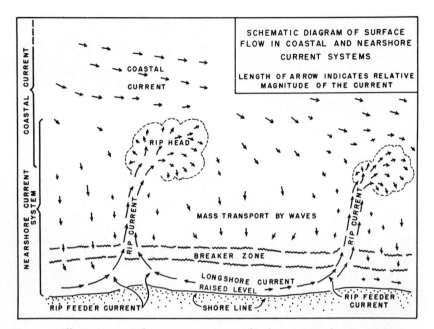

Figure 7. Illustrating typical current movements related to coastward transport by waves. Note the raised level along the shore in the center of the diagram and the currents moving out in either direction. The flows become much stronger in the rip currents but are diminished outside the breaker zone.

that flows out from the point of high level, moving along the shore in a direction that in this case is opposite to that from which the waves are approaching.

As a result of both of these types of longshore currents, considerable scouring out of holes is likely to occur just outside the beach. For this reason a bather wading out from the beach often encounters rather deep water near shore. If he traverses this deep trough-like area, he will usually find a bar out beyond, where the water is much shoaler. In some places the currents in the long-shore troughs are of such velocity and depth that they are dangerous to swimmers.

Rip currents. The water carried in by the waves must return, otherwise the water would keep getting higher and higher along the shore. A path is formed for the return at some low point in the longshore bar. This return flow, which is particularly pronounced along those beaches where there are large waves, develops the dangerous offshore movement known as a *rip current,* or more popularly as a *rip tide*[3] or a *sea puss* (Figs. 7, 8).

It is these return currents that account for most drownings attributed to undertow. The study of rip currents has shown no evidence that a swimmer is pulled under by the water flow, although he may sink due to swallowing water in his panicky and exhausted condition as he is buffeted by the agitated

[3] A misnomer, as they are not related to tides.

Figure 8. Aerial photograph from Oregon coast of a rip current emerging from the surf zone and forming a clockwise swirl. Currents like this endanger bathers because of their strong seaward flow. Photograph by J. D. Isaacs, Scripps Institution of Oceanography.

water surface. Along the shore the rip current extends from the surface to the bottom, but seaward it becomes largely a surface current. The lifeguards in southern California make almost all their rescues in rip currents, and the importance of these currents is now so well recognized that Coast Guard helicopters are sometimes used to advise or rescue bathers caught in a rip. The unwary bather may lose his balance in the trough where the current is flowing along the shore and then be swept out into the rip, or he may stumble directly into the channel produced by the outgoing rip itself. Sometimes the sides of these channels are quite steep and thus constitute a serious danger to a bather who is jumping up and down in the surf without realizing that he is being moved toward the rip by a current. Most rip currents are not a hazard to strong swimmers, although even they may encounter considerable difficulty in swimming against the current. The best means of escape is to swim parallel to the shore until free of the current on one side or the other, where the water is shoaler and where the current is either slack or moving toward the beach.

Rip currents have their useful aspects, however. Surfboard riders often look for a good rip to carry them out through the breakers, for in addition to the current the breakers here are usually smaller. Similarly, the rips can be used, although with caution, by expert boatmen in taking their skiffs out through the breakers. The deep water in the channels and the decrease in breaker height both are helpful.

It is often easy to recognize a rip when you are standing on the beach. You can generally see a place where the waves are not breaking as actively as they are on either side because of the deeper water. On the other hand, small short-period waves may break farther out, looking very much like the small breakers in the tide race at Hell Gate in East River, New York. The rip current has highly agitated water with small, slapping waves. Sometimes one can see the roiled-up water that is being carried out along the rip current, in contrast with cleaner water on either side. If you are standing on a little height above the beach, as in a lifeguard tower or on a bluff, you can see the color of the turbid water extending out along the course of the rip, and often there is concentration of foam, called a *foam line,* which develops along the boundary of one of the outward-surging masses in the rip.

The longshore currents and rip currents are important transporters of sediment, and during periods of large waves it is possible to find suspended sediment in rip currents thousands of feet out from shore. During the occasional rainstorms along the California coast, the muddy water that flows into the ocean is carried seaward by similar currents getting into the offshore circulation, which carries the sediment far from shore. The sharp boundary between muddy and clear water makes it possible to trace the flows for great distances from shore, sometimes a hundred miles or more.

Coral-reef rip currents. Currents that are similar in principle to rip currents are found among the channels of coral reefs. These currents are particularly dangerous because they are intermittent and between times of flow the area may look smooth and inviting. The reef, acting as a breakwater to the open ocean swell, may be topped from time to time by especially large waves that bring an excess of water into the channels inside the reef. This water must finally escape, but it may flow for a considerable distance along the inner channel before finding a gap in the reef through which it can return to the open sea. The gap may thus be returning the water that has flowed over the reef for a mile or more, and hence the return flow may move at a high speed. I once got caught in a current of this sort at Midway Island while collecting corals in a channel, and it was all I could do to hang on to a projecting pinnacle to avoid being swept out into the zone of breakers outside the reef. Fortunately the current was of short duration, and I was soon able to swim to a nearby reef.

Tides and Tidal Currents

The tides due to the moon and sun (which should not be confused with the badly named rip tides nor with tidal waves) are an important cause of currents that sweep the ocean bottom. Tides are extremely complex in their behavior. It is widely known that they are due to the gravitational pull of the moon and, to a smaller degree, to the sun acting on the ocean waters. Contrary to what might be supposed, however, the tide is rarely high when the moon is directly overhead. The effect of the moon alone on a globe entirely covered by water would be a rise in the water level when the moon was overhead, and an equal rise would occur on the other side of the world, for the moon would be pulling the water away from the earth on the side toward the moon and the earth away from the water on the other side. This is the same effect that comes from pulling a man by one arm. As he moves toward you, his other arm rises quite involuntarily, due to inertia.

The tides have the highest ranges when the moon and sun are either on the same side of the earth or on directly opposite sides. Under these conditions, which coincide with the new and the full moon, the tides are called *spring tides,* although this has no seasonal significance. When the moon is at right angles to the sun in relation to the earth, the two forces oppose each other, and the small tidal ranges that result are known as *neap tides.*

Various other factors influence the time of tides and the direction of their currents. The shores of the continents interfere with any orderly progression, and complex movements develop along the margins of the continents. Nodal points with virtually no tidal rise or fall develop in the center of ocean basins, whereas the tides progress as very long-period waves up and down the coasts of the continents. Long bays, like the Bay of Fundy or the English Channel, which have natural periods nearly in resonance with the tidal period, funnel the water in such a way as to produce enormous tidal ranges, especially at the bay heads. On the other hand, most inland seas, like the Gulf of Mexico and the Mediterranean, have very small tides because of their restricted entrances.

Tidal currents reach high velocities when there are two areas in close proximity that have at any one time large differences in tidal height. If there is a narrow water passage between these areas, the tidal flow will be greatly augmented. Thus bay entrances and the narrows between islands often have strong tidal currents. Passamaquoddy Bay on the Maine and New Brunswick coast is a well-known example of a place where the tide offers possibilities for developing power. Velocities of these currents run as high as fifteen knots, the highest on record being in Seymour Narrows, between Vancouver Island and the coast of British Columbia.[4] There the water moves at the speed of a mountain torrent and certainly must erode the bottom very actively.

One of the peculiar things about tidal currents is that, at least theoretically, the flow continues with little loss of velocity from the surface almost to the bottom, so that bottom erosion due to tides should be and probably is very pro-

---

[4] Now considerably reduced due to the dynamiting of a large rock mass that formerly constricted the channel.

nounced. Actually, however, there have been very few measurements of these bottom tidal currents. During World War II, I was on a ship anchored at the entrance to the Golden Gate, and we found with current meters that in 100 feet of water there was a six-knot current at the surface and a three-knot current a few feet above the bottom. This decrease is due to the friction of the bottom acting on the moving water mass. Deep entrances to bays usually result from such tidal currents. Thus the Golden Gate has a natural channel with one hole 360 feet deep, whereas one entrance to the Sea of Japan has a hole with depths of about 1,500 feet with shallow water on either side. In both cases the depth must be the result of the erosion of the tidal current moving in and out of the bay. Even where erosion does not develop at bay entrances, deposition is greatly impeded, so that the bottom is often rocky or may consist of coarse sediment, such as gravel. Muddy bottoms in narrows where the tide runs strong are found only if the currents encounter nothing but mud in their quarrying operations.

Slow-moving ships are much at the mercy of tidal currents. Unless a vessel is capable of high speed, it may be impossible for it to enter a narrows when the tide is running in the wrong direction. At other times the tidal currents may constitute a serious danger because of the possibility of an eddy washing a vessel onto the rocks. In the old days before the channel was adequately buoyed, one could have a harrowing experience going through Woods Hole near the Cape Cod Canal. I remember one such trip on a sailboat during a spring tide. We had to anchor at the west entrance of the Hole until the tide turned in our favor. By the time we got sails up and were underway, the favorable current had picked up such a velocity that the buoys were actually carried under water and could only be recognized by the swirling eddies around them, which did not differentiate them from the rocks. The channel was so crooked and the speed so great that it was almost impossible to keep our position on the chart. All we could do was to avoid all the larger eddies and hope that this would keep us from colliding with the rocks. We dodged back and forth, and after a few hectic moments we were through the narrows and sailing rapidly out into Vineyard Sound at the far end. The placid waters of the sound were a welcome sight.

Oceanic Currents

The Gulf Stream and other similar oceanic currents also affect the bottom. These currents, discussed much more completely in other books,[5] are set into motion by the two sets of prevailing winds, that is, the easterly trade winds

[5] See, for example, *The Sea Around Us* (New York: Oxford University Press, 1951), by Rachel Carson; and *The Ocean River* (New York: Charles Scribner's Sons, 1952), by Henry Chapin and F. G. Walton Smith.

in the tropics and the westerlies in higher latitudes. The result is the development of large gyrals, flowing constantly, although with speeds varying both in place and time. Separate currents flow around the ocean basins north and south of the equator, in a clockwise direction in the Northern Hemisphere and in a counterclockwise direction in the Southern Hemisphere because of the prevailing direction of the winds.

Many measurements have been made of the surface and mid-depth speeds of the Gulf Stream. Although few actual current-meter measurements have been made along the sea floor beneath these currents, recent photographs show that the bottom currents are flowing with sufficient velocity to produce current ripple marks. These currents move either north or south with velocities up to one knot. In 1885 J. E. Pillsbury on the United States Coast and Geodetic Survey steamer *Blake* measured currents in the Gulf Stream 2,500 feet below the surface at a height of 100 feet above the bottom. He found appreciable flows at this point. Bottom photographs indicate that strong Gulf Stream currents continue down to depths of almost a mile, where they sweep the bottom. The surface of the Blake Plateau, an area off the coast of Florida and Georgia, having depths of 2,500 to 3,500 feet, is apparently swept clean by the Gulf Stream. Sampling operations from this area have yielded mostly rocks and calcareous organisms that were growing on a hard surface. Apparently little fine sediment can lodge here. Along the coast of Japan, the Japanese Current (called also the Kuroshio), which has a somewhat lower velocity than the Gulf Stream, is underlain by a rocky bottom at a depth of at least 4,500 feet. This rock bottom occurs on a slope, and the rock may have been exposed as the result of landslides, whereas on the Blake Plateau off Florida there is an almost horizontal surface, which certainly would not have been affected by landslides.

The bottom photography of Lamont Geological Observatory has recently produced evidence indicating a swift southwesterly current flowing along the lower part of the continental rise off the northeastern United States, and lower on the rise a northwest flow is apparently related to the Gulf Stream. A ridge extending into the deep water south of Cape Lookout, North Carolina, has been attributed by Woods Hole geologists to sediment transported by these deep underflows and deposited where two currents have come together, resulting in a loss of velocity.

Turbidity Currents

If you are swimming under water with a face plate along a steep mud slope and you stir the bottom, you will develop on a small scale a type of current that is thought to be of great importance in the deep ocean. The mud you stir into the water will make the water heavier, and hence the mud and water will flow down the slope. Because of the turbid nature of these flows, they are usually referred to as *turbidity currents*. They were first discovered in reservoir lakes. Periodically these turbidity currents move down the great length of Lake Mead and stop only when they reach Hoover Dam. As far as is known, the currents in these lakes are very slow, and they carry only clay sediments. Using a tank, the well-known Dutch geologist Philip Kuenen demonstrated at Groningen University that coarse sediments, such as sand and gravel, can also be moved by currents of this sort. He calculated that high velocities might be attained if a large thick mass of saturated sediment started to slip on a submarine slope and became thrown into an aqueous suspension. The action of such masses has been compared to the violent flow produced by the deep water when a dam is broken and a lake moves en masse down the valley below the dam.

The possible existence of powerful turbidity currents on the ocean floor is suspected because of two lines of evidence. In the first place, sand layers and, much less commonly, gravel layers have been found by recent exploration in many places on the ocean bottom where the water is far too deep for the material to have been carried there by wind-wave action. Nor do these coarse sediments show any relation to the major ocean currents. Furthermore, they are definitely not confined to areas where there are strong surface currents. Many of the sand layers are found on great fans at the base of submarine slopes and near the mouths of submarine canyons. Therefore, it seems reasonable that slipping of material on slopes or along canyon axes has been involved in the cause of this phenomenon. The well-sorted, clean sands found in many of the layers and the graded bedding (changing progressively from coarse to fine, upward in the layers) suggest that currents rather than landslides have been responsible for the emplacement of the sediment. Furthermore, it is unlikely that a landslide could move across the very gentle slopes of the deep submarine fans, where so many of the sand layers are found.

The second line of evidence favoring powerful currents comes from the breaking of cables. The most striking incidence of this occurred south of the Grand Banks after the earthquake of 1929. Bruce Heezen and Maurice Ewing of Lamont Geological Observatory first called attention to the possible significance of the fact that these cables broke in an orderly sequence down the slope away from the earthquake epicenter (Fig. 9). They thought that these cable breaks were the result of great turbidity currents moving down the slopes. They postulated that the early stages of flow attained an average speed of 60 miles an hour because of the two-hour interval between breaks 120 miles apart. A re-evaluation of the data (Fig. 9) indicates that the speed, if due to turbidity currents, was not greater than 17 miles an hour, for cables broke instantaneously as far out as 100 miles from the epicenter. Other cable breaks in the Mediterranean led Heezen to infer equally rapid currents. In both cases, however, it is difficult to differentiate between breaks due to landslides and those due to turbidity currents, so that these excessively high speeds may not be justified.

Figure 9. Cable breaks at the time of the Grand Banks earthquake. These have been attributed to landslides, turbidity currents, or sudden readjustment of the packing of the sediments under the cables. The times of the breaks are indicated as far as they could be determined, although some of the second breaks occurred at undetermined times. Note that the times are given using the 24-hour-clock system and that the times of the outer breaks refer to the next day. The earthquake took place at 1531, and many of the breaks occurred almost immediately thereafter. If the breaks of the outer cables were due to turbidity currents starting in the area where so many cables broke at the time of the earthquake, the speed of travel must have been relatively high.

In an investigation of the subject, one of the leading authorities in soil mechanics, Karl Terzaghi of Harvard University has suggested as an alternative that a progressive and temporary liquefaction of slope sediments, moving like a wave down the slope after the earthquake, caused the cables to sink deeply into the temporarily liquefied slope and break as the result of distortion and stretching. This avoids the difficulty of explaining the enormous speed implied by the turbidity-current origin. The nature of the cable breaks is believed by Terzaghi to exclude the possibility of turbidity-current action. The resistance of the water also makes it seem unlikely that high velocities would develop.

Recent observations from deep-diving vehicles have shown that tractive currents, definitely not turbidity currents, move with considerable speed along the floors of submarine canyons. The finding that the character of many deep-sea sands could be more easily explained as due to these tractive currents has also somewhat weakened the case for turbidity currents.

In any case, there are indications that turbidity currents are capable of transporting coarse sediment out into very deep water, and it may well be that they are also important agents in eroding the ocean floor. Their importance will be discussed further in Chapter VII, which deals with the origin of submarine canyons.

Internal Waves

Despite the comparatively small number of photographs of the deep-ocean floor, some of them have been found to contain definite indications of the operation of relatively strong currents. For example, ripple marks were photographed at a depth of 4,500 feet on a flat-topped seamount (Fig. 10). Probably the top of this mountain was once in the surf zone millions of years ago, but the ripples produced at that time could not have lasted. This same photograph includes trails left by organisms, which must quickly destroy such ephemeral features as ripples. Therefore, the ripples must represent a very recent development, although the foraminiferal sand is of low density so that it is rippled more easily than mineral sand. Even more surprising is the finding of scour marks around nodules on the floor of the deep Altantic (Fig. 11). These also would soon be eliminated either by organisms or by deposition. It is unlikely that turbidity currents have produced these ripples, since in neither case was there a slope down which such currents could have moved.

Internal waves may account for some of the deep ripples. Waves of this type operate within a body of water without having any effect on the surface. The principle can be demonstrated by putting two liquids of different density and color in a glass tank so that a color boundary can be seen through the side plate. Then agitate the tank very gently by moving the ends. Now looking through the glass you can observe a series of waves moving up and down at the line of contact between the different colored liquids. These waves will have far greater amplitude than any waves formed at the surface of the tank. Internal waves in the ocean were first described by the Swedish oceanographer Otto Pettersson as the result of his observations on the so-called dead water that is observed

Figure 10. Ripple marks in foraminiferal sand at a depth of 4,500 feet on Sylvania Seamount, Marshall Island area. Photograph by Carl J. Shipek, Navy Electronics Laboratory.

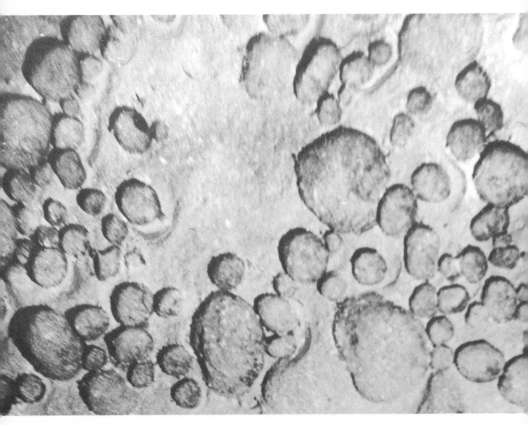

Figure 11. Photograph taken at 18,000 feet in basin 300 miles east-southeast of Bermuda. Shows current scour next to round objects, which are probably manganese oxide nodules. Photograph by David Owen of Hudson Laboratory, with Ewing deep-sea camera.

under the keels of vessels in the Arctic. Here fresh water from melting ice overlies the salt water of the ocean.

Internal waves were later found by repetition of temperature readings in the deep ocean at several definite levels. Since the temperature normally decreases progressively with depth, a large internal wave will cause an alternate rise and fall of the temperature within the body of the oceans. Little is known about the velocity of the bottom currents produced by these curious waves of the ocean interior, but theoretical computations indicate that they may be sufficiently powerful to disturb the sea floor and even to produce ripple marks.

There are other types of waves and currents in the ocean, many of them still too poorly understood to be discussed here. The badly named tidal waves, which have no relation to the lunar or solar tides, are usually the result of movements on the sea bottom accompanying and, in fact, causing earthquakes.

These waves, which scientists refer to as tsunamis (taken from a Japanese word), or seismic sea waves, have such long wave lengths that theoretically they can affect the entire ocean bottom. Because of special problems that they raise and because of their special interest to man, they will be considered at length in the next chapter.

From the discussion of the various types of water movements in the ocean, we have found that they extend to much greater depths than was previously supposed and that they may have played an important role in the origin of the features of the deep-sea floor. On the other hand, our evidence from actual measurement of bottom currents is still rather inadequate. Various devices that have been contrived recently should produce important progress in the near future. The indirect evidence now available, partly photographic, indicates that currents may become a major factor in our hypotheses.

CHAPTER II

CATASTROPHIC WAVES FROM THE SEA

Experiences in the Hawaiian Disaster of April 1, 1946

The term *tidal wave* has had an ominous sound in Hawaii since April 1, 1946. My own experience on that day may serve to introduce the discussion of tidal waves, or tsunamis. At that time my wife and I were living in a rented cottage at Kawela Bay on northern Oahu (Fig. 12). On the previous day, a

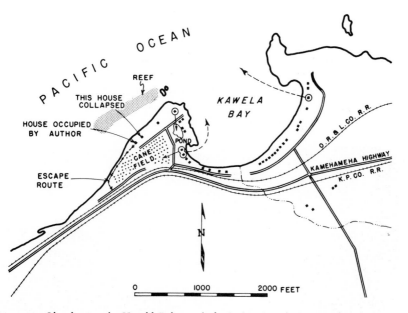

Figure 12. Sketch map, by Harold Palmer of the University of Hawaii, showing area in which the author experienced a tsunami.

Sunday, the beaches and reefs were swarming with people and the cottages alive with activity. Fortunately, almost everybody left to go back to Honolulu that night. Early the next morning we were sleeping peacefully when we were awakened by a loud hissing sound, which sounded for all the world as if dozens of locomotives were blowing off steam directly outside our house. Puzzled, we jumped up and rushed to the front window. Where there had been a beach previously, we saw nothing but boiling water, which was sweeping over the ten-foot top of the beach ridge and coming directly at the house. I rushed and grabbed my camera, forgetting such incidentals as clothes, glasses, watch, and pocketbook. As I opened the door I noticed with some regret that the water was not advancing any farther but, instead, was retreating rapidly down the slope.

By that time I was conscious of the fact that we might be experiencing a tsunami. My suspicions became confirmed as the water moved swiftly seaward, and the sea level dropped a score of feet, leaving the coral reefs in front of the house exposed to view. Fish were flapping and jumping up and down where they had been stranded by the retreating waves. Quickly taking a couple of photographs, in my confusion I accidentally made a double exposure of the bare reef. Trying to show my erudition, I said to my wife. "There will be another wave, but it won't be as exciting as the one that awakened us. Too bad I couldn't get a photograph of the first one."

Was I mistaken? In a few minutes as I stood at the edge of the beach ridge in front of the house, I could see the water beginning to rise and swell up around the outer edges of the exposed reef; it built higher and higher and then came racing forward with amazing velocity. "Now," I said, "here is a good chance for a picture." I took one, but my hand was rather unsteady that time. As the water continued to advance I shot another one, fortunately a little better (Fig. 13). As it piled up in front of me, I began to wonder whether this wave was really going to be smaller than the preceding one. I called to my

Figure 13. The advance of the tsunami onto the beach ridge where the author was living on April 1, 1946. A few seconds after this was taken, the wave swept over the area in the foreground, bringing in large coral boulders and causing great property damage.

wife to run to the back of the house for protection, but she had already started, and I followed her just in time. As I looked back I saw the water surging over the spot where I had been standing a moment before. Suddenly we heard the terrible smashing of glass at the front of the house. The refrigerator passed us on the left side moving upright out into the cane field. On the right came a wall of water sweeping toward us down the road that was our escape route from the area. We were also startled to see that there was nothing but kindling wood left of what had been the nearby house to the east. Finally, the water stopped coming on and we were left on a small island, protected by the un-damaged portion of the house, which, thanks to its good construction and to the protecting ironwood trees, still withstood the blows. The water had rushed on into the cane field and spent its fury.

My confidence about the waves getting smaller was rapidly vanishing. Having noted that there was a fair interval before the second invasion (actually fifteen minutes as we found out later), we started running along the emerging beach ridge in the only direction in which we could get to the slightly elevated main road. As we ran, we found some very wet and frightened Hawaiian women standing wringing their hands and wondering what to do. With difficulty we persuaded them to come with us along the ridge to a place where there was a break in the cane field. As we hurried through this break, another huge wave came rolling in over the reef and broke with shuddering force against the small escarpment at the top of the beach. Then, rising as a monstrous wall of water, it swept on after us, flattening the cane field with a terrifying sound. We reached the comparative safety of the elevated road just ahead of the wave.

There, in a motley array of costumes, various other refugees were gathered. One couple had been cooking their breakfast when all of a sudden the first wave came in, lifted their house right off its foundation, and carried it several hundred feet into the cane field where it set it down so gently that their break-fast just kept right on cooking. Needless to say, they did not stay to enjoy the meal. Another couple had escaped with difficulty from their collapsing house.

We walked along the road until we could see nearby Kawela Bay, and from there we watched several more waves roar on to the shore. They came with a steep front like the tidal bore I had seen move up the Bay of Fundy at Moncton, New Brunswick and up the channels on the tide flat at Mont-Saint-Michel in Normandy. We could see various ruined houses, some of them completely demolished. One house had been thrown into a pond right on top of another (Fig. 14). Another was still floating out in the bay.

Finally, after about six waves had moved in, each one apparently getting progressively weaker, I decided I had better go back and see what I could rescue from what was left of the house where we had been living. After all, we were in scanty attire and required clothes. I had just reached the door when I became conscious that a very powerful mass of water was bearing down on the place. This time there simply was no island in back of the house during the

Figure 14. Showing two houses washed into a small lake at Kawela Bay, Oahu, during the 1946 tsunami. Photograph by U.S. Navy.

height of the wave. I rushed to a nearby tree and climbed it as fast as possible and then hung on for dear life as I swayed back and forth under the impact of the wave. Like the others, this wave soon subsided, and the series of waves that followed were all minor in comparison.

After the excitement was over, we found half of the house still standing and began picking up our belongings. I chased all over the cane fields trying to find books and notes that had been strewn there by the angry waves. We did, finally, discover our glasses undamaged,[1] buried deep in the sand and debris covering the floor. My waterproof wristwatch was found under the house by the owner a week later.

"Well," I thought, "you're a pretty poor oceanographer not to know that tsunamis increase in size with each new wave." As soon as possible I began to look over the literature, and I felt a little better when I could not find any information to the effect that successive waves increase in size, and yet what could be a more important point to remember? You can be sure that since then those of us who have investigated these waves in the Hawaiian Islands have stressed this danger, and I was most happy to find recently at a local island store a tidal-wave warning that emphasized the crescendo to be anticipated in future disasters. Nowadays, also, there are tidal-wave warning alarms that send out alerts either when reports of earthquakes under the ocean indicate dangerous possibilities, or when early waves arrive at other islands along the general route, or when the tide begins to fluctuate in an abnormal fashion. The importance of these warnings can be seen when it is noted that most of the 159 people who were lost during the 1946 tsunami could have saved their lives by running from the scene to higher ground when the waves first began. The Hawaiians are early risers, and being always attuned to the varying moods of the ocean, almost everyone was conscious of a sudden diminution of the noise of the breakers when the sea withdrew. Most people ran to see the strange sight of the reefs being laid bare, and many went out on the reefs to pick up the stranded fish. The 1957 tsunami was almost as destructive to property in Hawaii as that of 1946, but thanks to the warning system no lives were lost. I was shocked to learn that another house in which I had vacationed was destroyed by the 1957 waves. In May, 1960, the warnings were ignored, and the waves from the great Chilean earthquake took 61 lives at Hilo.

Tsunamis and Their Significance

The meaning of the Japanese word *tunami* (pronounced 'tsunami' and hence written that way) is 'large waves in harbors,' a good name, as it takes a disturbance of this kind to produce large waves in sheltered bays. The tsunamis certainly do not have anything to do with the tide, although the approach of the waves on an open coast where there are no reefs looks like a rapid rise of the tide, hence tidal wave.

Most tsunamis apparently have their origin in the great sea trenches that surround the margin of the Pacific Ocean (see Chap. VIII). Fortunately for those who live on the west coast of the United States, there are no deep trenches in this section, which is perhaps the reason no appreciable tsunami has been observed along the California coast. Almost all tsunamis are preceded

[1] The sand, being coral, did not produce scratches.

by world-shaking earthquakes, in which all seismograph stations have recorded the earth tremors from the disturbance. It seems likely that the waves are caused by faulting, a sudden dropping or lifting of a segment of the ocean bottom, which results in a displacement of large amounts of water. An alternative explanation is that huge submarine landslides produce the waves, although there is no good confirmation of this idea, and all tsunamis except those caused by volcanic eruptions have followed large earthquakes. If the ocean bottom drops, the surface waters are sucked into the hole, and when the water flowing from either side comes together, the surface of the water rises and waves move out in all directions under the force of gravity. Alternatively, if the bottom rises, the water is lifted and moves outward.

The waves are most violent in their effect in a direction at right angles to the fault. Since the Aleutian Trench, south of the islands of that name, runs east and west, movement along the faults that bound the trench produces waves that are most significant to the north and south, as were those of 1946 and 1957. Almost no one lives along the south exposed side of the Aleutian Islands, so that little damage has resulted in that area, although in 1946 the water rose at Scotch Cap on Unimak Island to over 100 feet, destroying a lighthouse and flowing over a 100-foot terrace. The Hawaiian Island group, more than 2,000 miles to the south, had waves that washed up to a maximum height of 57 feet, as far as we were able to determine. Fortunately, in most places it did not rise nearly as high as that (Fig. 15).

Tsunamis move at an enormous speed in the open ocean, averaging about 450 miles an hour. This varies directly with the depth of the water, because these waves, unlike wind waves, have very long periods, commonly fifteen minutes, and have distances of as much as 100 miles between crests. Substituting these values in Figure 3 will explain why the depth of the water influences the speed of the waves. Their height in the open ocean is so small, however, that they may have no erosive effect on the deep-sea floor. It is only along a coast that they become destructive.

The waves took about four hours to reach Hawaiian shores after the Aleutian earthquake of 1946. As the waves came into shallow water, they were greatly slowed down, so that they advanced at a rate of only about 15 miles an hour as they approached the coast. As their energy became confined to shallow water, they grew in height. The exposed coasts on the north of the Hawaiian Islands had large waves, whereas small heights were observed on the protected south side of the islands.

The investigations that followed the tsunami resulted in some conclusions that may prove helpful in ameliorating the effects of future calamities of this sort.[2] The increasing height of the successive waves was perhaps the most im-

---

[2] For details see "The Tsunami of April 1, 1946," by F. P. Shepard, G. A. Macdonald, and D. C. Cox, *Bull. Scripps Inst. Oceanog.*, 5, No. 6 (1950), 391-470.

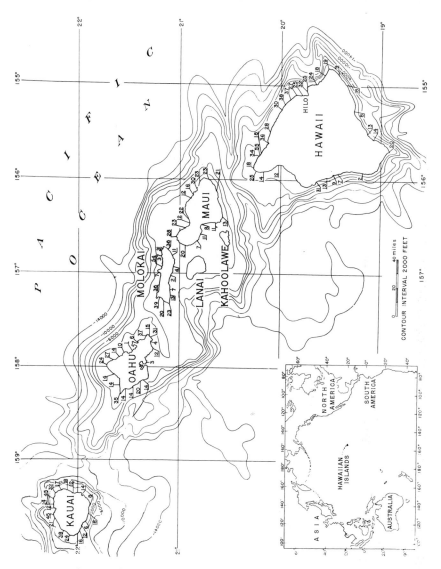

Figure 15. Showing the height in feet to which the waves rose on the shores of the Hawaiian Islands on April 1, 1946.

portant lesson. We found that in some places the second or third waves were the largest, but elsewhere the seventh or eighth reached the greatest height. On the western coast of Hawaii (the big island) some waves actually came in during the following night after an interval of eighteen hours and reached heights greater than those experienced that morning. These surprising reports were confirmed by a considerable number of sources, but are not readily understood. It can only be supposed that the waves represented a reflection from a submarine cliff off Japan and another reflection from an escarpment in Oceania, so that finally the waves, after making what is comparable to a three-cushion shot in billiards, arrived at their destination. In any case the danger of possible late wave arrivals, especially on protected sides of islands, cannot be minimized.

About the most dangerous thing that a person can do during a tsunami is to walk out on the exposed reefs to gather up the fish left by the retreating seas. In 1946 many of the drownings in Hawaii occurred as a result of this activity, a natural reaction of people whose livelihood comes from the sea and who for the most part had never even heard of a tsunami (the last one of any size having occurred in 1877). The building of sea walls in front of a town, as had been done at Hilo, is helpful even if the sea wall is knocked over by the advancing waves. Undoubtedly the friction considerably decreases the power of the waves. Wherever possible the restriction of building to zones that have at least moderate elevation above sea level in danger areas is recommended.

In the tropics the corals have been very helpful in sparing man from even worse trouble from tsunamis by building large protective reefs along many coasts. The widest reef in the Hawaiian Islands, at Kaneohe Bay, is found on the north side of the Island of Oahu and therefore on the side from which the waves approached. Yet, this wide reef seems to have been entirely effective in stopping the progress of the waves. Most people living in its lee were not even aware that a tsunami had occurred. Heights of not more than one or two feet were all that could be found by careful investigations along this shore. Other areas where reefs had smaller widths were less fortunate, as at Kawela Bay, where we were living behind a small reef and where the water rose ten to nineteen feet. However, the height of the raised water level, behind these reefs, was in almost every case less than in adjacent areas, where the water came in unimpeded.

Similarly, the existence of a submarine valley or canyon off a coast definitely has an important effect. Just as ordinary wind waves are small at the heads of submarine valleys (Fig. 4), so also tsunamis are greatly reduced by the spreading of the energy as the waves move up the valleys at a faster rate than over the intervening ridges. Conversely, the waves traveling over a submarine ridge are particularly large. For example, three ridges extend down the slope on the north side of Kauai Island, and over these the waves attained their greatest heights (Fig. 15). So if you live in Hawaii and want to live next to the beach, build your house behind a coral reef or look at a chart to see if there is a submarine valley out in front.

A few destructive tsunamis have occurred in localities where no great trench is known to exist. Among these are the waves that swept in on Lisbon in 1755, moving up the Tagus River and causing a very heavy loss of life. These followed a great earthquake with a center under the Atlantic some distance off the shore.

The most destructive waves of all time have been related to volcanic activity. In 1883 when Krakatoa blew off its head, a sudden engulfment occurred, setting up very unusual waves. These rolled in on the adjacent islands of Java and Sumatra and drowned tens of thousands of natives, rising, it is said, to heights of well over a hundred feet. Curious reports came from these waves. They showed on the tide gauges all the way around the world, even in the English Channel. If these tide-gauge records indicated a tsunami actually coming from the Krakatoa engulfment, the waves must have been reflected from numerous submarine escarpments in order to have reached such a destination. The waves were recorded also in the Hawaiian Islands, although here the time of arrival does not agree with the time that one would predict for a wave traveling from Krakatoa to the Hawaiian Islands. The explanation is still in doubt.

It is disturbing to consider what would happen if a tsunami should come into a shore like Long Island, where some of the beaches have hundreds of thousands of bathers during a warm summer day. We have no records of dangerous waves coming in at these places. However, in the case of the 1929 Grand Banks earthquake, which wrecked a large part of the submarine cables going between our east coast cities and Europe, there had been no previous record of tsunamis in the area. The waves accompanying the Grand Banks earthquake moved in on Burin Peninsula on the south coast of Newfoundland, rising to fifteen feet. Such rises would, of course, sweep over most of the beaches along the exposed portion of the east coast. Let us hope that no new submarine faults come suddenly into being and send waves into this area. The effect would be almost as bad as that of a hydrogen bomb.

A Swash Seventeen Hundred Feet High

If you have a bathtub with an inclined end and you have children, you probably know what happens when the children slide down the incline into the tub. The water swashes up onto the other end and may pour out onto the floor. Such *swashes* occur also in nature. They are particularly common in fiords and steep-walled lakes, where they are produced by great rock falls coming down from the mountainsides and pushing the water across the fiord or lake so that it rises onto the farther side. At Loen Lake in Norway tourists are shown the remains of a boat that was carried a hundred feet above the lake level by such a swash.

In July 1958 one of the largest of all these waves took place in Lituya Bay along the south coast of Alaska. This is an uninhabited location, but three fishing boats were located in the lower end of the bay. There was a great earthquake as a result of movement along a fault at the head of the bay, six miles from the location of the boats. The boatmen observed the commotion and a few minutes later saw a great wave coming down the bay toward them. One boat was lifted by the front of the wave right over the spit at the entrance of the bay, and the occupants say they were lifted so high they could look down on the eight-foot fir trees growing on the spit. They were dropped stern first and the boat sank, but they escaped in a skiff. Another boat was at anchor and, being unable to get up the anchor, headed toward the approaching wave. They were lifted by it, and the anchor cable before parting stopped them from being carried over the spit, so that their boat was saved from destruction. The third boat just disappeared.

Examining the sides of the bay after the earthquake, geologist Don J. Miller of the United States Geological Survey and seismologist Don Tocher of the University of California found that the wave had swept along the side of the bay at about the 100-foot level, knocking down virtually all of the trees and stripping them of their bark. The greatest surprise was what happened on a mountain spur across a narrow inlet from the largest of the rock falls. Here a swash swept up onto the ridge to a height of 1,700 feet, making a clean sweep of the forest so that only one tree was left standing amidst the ruins. The trees at the upper level were washed into the living forest, showing that it was a wave rather than a landslide that caused this damage. Miller, flying along the bay a few hours after the quake, saw great masses of water still running down the sides of the mountain. On the more distant side of the spur the trees were tipped along the slope away from the advancing water, giving further evidence that it was a great swash that reached this unparalleled height above sea level.

Storm Surges

The term *tidal wave* has sometimes been used also to describe a rise in sea level that accompanies a hurricane. A *storm surge* seems to be a more acceptable term. In 1900 the sea rose about fifteen feet at Galveston, Texas, and topped the sea wall, sweeping into the city and drowning 6,000 people. A sea wall has now been constructed that will probably prevent any recurrence of this sort, but other cities are less well protected and subject to dangerous waves. In 1938 a great hurricane moved up the east coast, quite contrary to the predictions that had been made by the meteorologists, and passed inland across Long Island. The sea here rose also about fifteen feet, killing 600 people and causing tremendous amounts of damage to the beach property. It developed numerous new inlets in the beaches and changed the appearance of the coast until it was practically unrecognizable after the waves had stopped. Several other hurricanes, notably those of 1954 on the east coast and that of 1957 in western Louisiana, have produced similar inundations. A far worse catastrophe of this sort occurred at the head of the Bay of Bengal in 1737, when 300,000 people were drowned during a hurricane. The rises accompanying these great storm waves are similar to tsunamis, except that the waves do not come in rhythmic succession. The rise may be quite as rapid, but the high water usually lasts a longer time, and recurrences are not particularly pronounced.

The damage caused by all of these rises of sea level is related only in part to the high water. In addition, the natural barriers to wind waves that exist along the shore and many of the artificial walls and jetties become less protective, so that the wind waves are superimposed upon the top of the sea-level rise and wreak their havoc at a new high level.

CHAPTER III

OUR TRANSIENT BEACHES

If you are a resident of the beach front north of the Redondo Municipal Pier in southern California, you know what the title of this chapter means because the broad beach that you once admired from your front windows has now disappeared and your property has become part of the state tidelands. The biblical saying "Build not thy house upon the sands" should have been heeded by many an owner of beach property. A lot may look fine in summer when the waves are small and the real estate agents are plying their wares with the promise that the property will soon double in value, but wait for those winter storms and see what happens to that new development. It just may not be there. You should be concerned also about the jetties the city is constructing along the coast near your beach. It is a fine thing to have a boat harbor, but what will happen to your beach? It may disappear after the jetty is built.

Beach Terminology

Scientists now have learned some of the reasons for the transient nature of beaches. Prior to discussion of the causes, however, it may be helpful to consider briefly the geological, engineering, and legal terminology that is applied to beaches (Fig. 16, Sec. A). The *beach* extends landward to the farthest point where sand has been transported by the waves.[1] On the landward side of the beach there may be a sea cliff or dunes, or the beach may terminate against a sea wall or some other work of man. The inner portion of a beach, wherever it is essentially horizontal or slopes landward, is called the *berm* or the *backshore*. The seaward-sloping portion of the beach is referred to as the *foreshore*. This extends to the low-tide level. Beyond this point the continuously submerged portion is referred to as the *offshore*. The backshore, which is above the high-tide line, may include several berms, and the foreshore may not slope continuously but may have a *low-tide terrace* that is as horizontal as the berm. The offshore frequently has troughs and bars, as indicated in Chapter I.

Other terms that are related to beaches and will be used in this book include *barriers,* which are sand bodies extending along the shore as *barrier beaches* if narrow or *barrier islands* if wide. The water bodies inside barriers are known as *lagoons* if they extend roughly parallel to the mainland shore. If one end of a barrier beach is connected to the mainland, it is called a *spit*. If

---

[1] Catastrophic waves such as those discussed in Chapter II are excepted in this definition.

a sand mass projects from the shore line as a point due to the connection of two converging spits, as at Cape Hatteras, the beach is called a *cuspate foreland*.

Beach Cycles

In all areas where storm seasons occur, the beaches undergo an annual cycle. During the periods when the waves are small (ordinarily in summer) the uprush of the waves is still sufficiently powerful to transport sand to the upper portion of the foreshore, but the backwash from the returning wave does not have enough power to return a large part of the sand carried landward in this way. As a result, the sand builds up and the berm is extended seaward (Fig. 16,

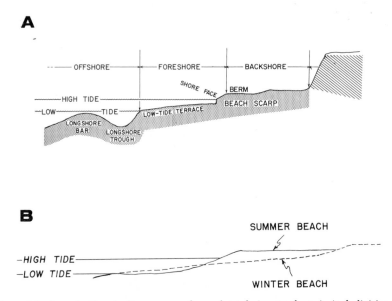

Figure 16. Sec. A: Terminology commonly used to designate the principal divisions of beaches. From Shepard, Submarine Geology, 2d ed. Sec. B: Typical contrasts in profiles between summer beaches with broad berms and winter beaches cut back by storms.

Sec. B). The long beach on the north side of La Jolla, California, for example, has a berm growth of several hundred feet during the summer (Fig. 17). In other localities if there is a net growth for a number of years, the berm not only becomes very wide, but the inner part is converted into a mass of small dunes, which may become covered with dune grass, so that the area becomes a semipermanent part of the land and loses its contact with the ocean.

Figure 17. Comparative photographs showing the same beach in summer, with a broad berm piled high by the sand, and in winter, cut away by storms exposing gravel that has formed cusps.

Now let us consider the stormy periods (usually in winter) when the waves are much larger than average. During these times the large waves pound against the front of the berm and stir great masses of the sand into suspension. The powerful backwash carries most of this sand down the slope. Since long-shore currents attain high velocities during periods of large waves, the sand that is torn from the berm is carried along the shore by these currents until it comes to a rip current, where it is carried seaward. The outward movement along the rip channels is far stronger than the return flow, which is diffused over a wider area. Therefore, the sand has a net seaward motion. As a result, the beach berm may be cut back rapidly, often with the development of steep scarps that are short lived.

Out beyond the beach, a trough reaching a depth of ten to twenty feet is commonly formed during heavy surf, and, seaward of the trough, submerged sand bars are built up from the bottom. The larger the waves, the farther out the bars will move, and hence the deeper will become the water on top of them. Conversely, when the waves decrease in size, the bars move landward and develop shoaler crests. These bars and troughs are of considerable importance in landing operations. In fact, our intensive wartime study of the bars near Scripps Institution was made because of the plans for landings along the beaches of Japan, where such bars were known to exist. A landing craft might be able to cross the deep bars during the calm intervals of the stormy season, whereas it would be grounded on one of the shallow summer bars under the same wave conditions. On the other hand, the winter trough may have currents that endanger the landing of personnel who are encumbered with heavy equipment. In summer the bars may be largely eliminated or move up entirely onto the low-tide terrace, so that they do not offer an obstacle and there is no dangerous trough.

As a result of changing wave conditions beaches go through many cycles. A short-period cycle develops in response to a single storm. Longer cycles are related to the storminess of winter seasons and the more peaceful conditions of summer. Nature seems to have timed this yearly cycle to fit the pleasures of man. Thus, expanded beaches are provided in the summer, when they can be used for relaxation, while the narrow beaches of the winter are of little detriment to man's enjoyment.

Beach changes may be greatly complicated by the direction in which the waves approach. We have a small beach at La Jolla called Boomer Beach because of the pounding of the waves on the submerged rocks immediately beyond the beach. Here the rocks make it a dangerous place to swim but a wonderful place for expert body surfers. If you visit this beach in summer, you will find the sand piled high along the shore, but in the fall after the pounding waves of the first northwest blow have hit the area, that sand disappears almost to the last grain, exposing a mass of boulders. Look to the south, however, and you will find that just the opposite has happened. In winter, sand has been piled on top of the rocks that were bare during the summer. This shifting of sand back and forth along the shore is a result of the change in the direction of wave approach. The northwest storms of winter move sand to the south and thus cover the southern end, whereas during the summer the long rollers that come into the area from the winter storms of the Southern Hemisphere approach from the south, so that sand moves northward. This is an important thing to bear in mind if you buy beach property. Just because you look at the property during the stormy season, do not be sure that this is the time when the beach is most denuded. You should learn something about the directions of wave approach in the area before you make your decision. La Jolla's Boomer Beach is by no means an isolated case; numerous beaches around the world are controlled by shifting of sand in response to changing directions of wave approach.

Jetties and Other Works of Man

The works of man have greatly altered beach histories. The simplest case is of a jetty having been built out into the sea, or of a pair of jetties having been constructed to form a harbor entrance. If the prevailing wave approach, or the approach of the principal large waves, is from the north, as it is along most of the California coast, where many problems of this kind have arisen, the sand that drifts southward along the shore will accumulate on the north side of the north jetty. Thus a wide beach is built to the north of the jetty. That is fine, unless the beach gets so wide that it becomes too much of a walk to attract swimmers. Downcurrent from the jetty, a different situation develops. The normal supply of sand that is brought into the beach during periods of low waves is no longer available, since it has been trapped behind the jetty. The net result is that downcurrent from the jetty the beach is cut back as usual during the winter storms, but during the ensuing summer the rebuilding is extremely limited. Sand that has been cut away in winter has in the meantime shifted with the current and hence is out of range. Therefore it is possible not only that the beach will disappear, but also that the adjacent shore property with houses or roads will be undermined.

The history of southern California is replete with examples of beach ruination. Along the east-west-trending coast at Santa Barbara, jetties were built (Fig. 18), and a wide beach developed to the west.[2] The beaches to the east were gradually cut away, so that millions of dollars have had to be spent to correct this situation. At Santa Monica (near Los Angeles), a jetty was built parallel to the coast (Fig. 19) with the hope that the sand, instead of being trapped on the upcurrent side, would move through the harbor and thus be

---

[2] The current runs from west to east at this place.

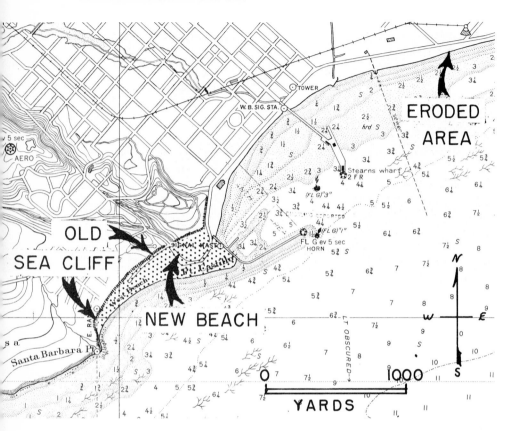

Figure 18. Changes in shoreline at Santa Barbara resulting from the building of the harbor jetties. Since this chart was made, the fill has extended around the west jetty and into the harbor, so that dredging has to be done to keep the harbor open. From U.S. Coast and Geodetic Survey Chart 5261. Soundings are in fathoms below mean low water level.

available for the beaches on the downcurrent side. However, the parallel jetty produced what is called a *wave shadow,* that is, it prevented the breaking of the waves between it and the shore. Without the turbulence in breakers it became impossible to keep the sand moving along the shore. Once the sand came to rest in the area protected by the parallel jetty, it built an unnecessarily wide beach at that point, filling much of the harbor. An added result was the cutting away of the beach on the downcurrent side.

At Redondo, along the southern part of Santa Monica Bay, the harbor jetty has had a particularly disastrous effect. It was built at the point where the waves are reinforced by wave convergence on the side of a submarine canyon (see Fig. 4). After the jetty was finished, serious coast erosion began despite the protection from the jetty. The difficulty was due to the continuation of the

Figure 19. Showing the results of building a breakwater parallel to the shore at Santa Monica, California. Note the wide beach that is threatening to fill the harbor. The beaches to the south (right) are extremely narrow. Foam lines streaming into the harbor result from waves breaking over the low breakwater. Photograph courtesy of D. L. Inman, Scripps Institution of Oceanography.

wave convergence, which cut away sand as it had before. Previously, however, sand brought into the area from the north had compensated for that which was lost by the erosion. Now this sand supply became trapped behind the jetty. The result was first the cutting away of the beach, and then the removal of several city blocks of valuable property. The situation has now been stabilized by extending the jetty toward the canyon and stopping the wave convergence.

Farther south, at Oceanside, the recession of the beaches started shortly after World War II. During the war the Marine Corps needed to build jetties just north of Oceanside. These jetties stopped the normal southward flow of sand along the shore, and therefore the beaches in the lee gradually became starved for sand. The Camp Pendleton jetties served little purpose because they formed only a temporary harbor. After the sand filled in along the north jetty, it entered the harbor, causing it to become so shoal that it was no longer of much use for boats. The size of the harbor was not nearly sufficient to develop the tidal flow that keeps many large harbors open. A similar effect developed in the harbor entrance at Mission Bay, north of San Diego (Fig. 20). Here again a wide beach grew out on the north side of the jetties, and the beach to the south suffered erosion despite a large addition of sand to the beach pumped in from the flood-control channel.[3] The sand building around the end of the

[3] Dredged by Army Engineers to take care of possible flood water from the intermittent San Diego River.

Figure 20. Aerial photograph of the jetties at the entrance to Mission Bay in 1953 prior to the cutting of the present twenty-foot channel. A ten-foot channel had been dredged, but transportation of sand soon built a bar at the entrance, so that the channel became impassable. Photograph by John MacFall of Scripps Institution of Oceanography for a survey by J. D. Frautschy and D. L. Inman, also of Scripps.

north jetty finally entered the channel so that it became shoaled sufficiently to be dangerous even for small boats. Eleven drownings resulted from the capsizing of boats in breakers during conditions of large waves. Now the harbor entrance has been dredged to a depth of twenty feet so that it is useful again, but considerable upkeep will be necessary to keep it open. Here, as at Camp Pendleton, the tide does not flow through the entrance with sufficient force.

Not all artificial jetties cause harmful results. Most jetties produce at least a minor adjustment to the shore line because their sharp angularity sets up unusual eddies. Thus the building of San Diego harbor jetties at the beginning of the century caused Coronado Strand to be temporarily eroded near the famous old Coronado Hotel (Fig. 21). In this particular locality, however, the jetty did not have a ruinous effect on the beach farther south, because sand in this area is largely supplied from the intermittent Tia Juana River, which enters the ocean to the south. The northwest movement of sand in this locality is in contrast to much of the rest of the southern California coast, where the northwest winds produce a southerly shifting of sands. The difference here comes from the protection from northwest storms provided by Point Loma. Waves from these storms are considerably decreased by having to bend around the southward-projecting point before they approach the shore. Thus, there is

Figure 21. Showing the destruction of the beach and of the road to the north of the Coronado Hotel, as a result of the local adjustment following the building of the San Diego breakwater.

only rarely southerly transport of sand along Coronado Strand. The large southerly swell that comes in from the Southern Hemisphere during the summer, on the other hand, takes the sand supplied by the Tia Juana River and moves it northward. This has resulted in general building out of beaches north from the Tia Juana River mouth.

The jetties at the well-known yachting harbor of Newport, California, south of Long Beach, have had only minor local effects on the beaches. This is also due, at least in part, to the sheltering from the northwest winds produced by Point Fermin and the Palos Verdes hills. In the case of Newport there has been more deposition on the northwest side of the jetties than on the southwest because the main source of sand is from the Santa Ana River, which enters the coast northwest of Newport.

The long straight beaches of Florida are interrupted by a number of jetties built to allow the development of the boat harbors for winter-resort activities. Here the problem has been somewhat simplified because of a double source of sand supply. One important source is the sands carried into the ocean by the rivers to the north of the Florida peninsula. This is quartz sand, and the northwest storms gradually move it to the south. This movement is assisted to quite an extent by a southerly current that develops as a return eddy from the northward-flowing Gulf Stream. Broad quartz sand beaches are formed on the north side of jetties, as at Daytona Beach, where auto races have been held for fifty years. To the south of the jetties, the Florida beaches consist predominantly of small shells known as coquina and include an abundance of the unicelled foraminifera[4] and various other marine organisms. The source of these shell sands is the organisms that live plentifully near the shore in the

---

[4] Foraminifera are very small one-celled animals, protozoa, mostly with calcareous shells. They either live on the bottom, benthonic, or drift with the ocean currents, planktonic.

warm coastal water. The waves carry in many of the shells after the death of the organisms and thus may maintain the beaches despite the effects of the jetties, which have robbed the downcurrent beaches of much of the supply of quartz sand that they formerly received.

Many important beaches the world over are partially protected by short jetties known as groins. Good examples are found along the coast of New Jersey (Fig. 22). These cause a slight widening directly upcurrent from the groin and a corresponding narrowing downcurrent, but have a net effect of slowing the transport of sand along the groined portion of the beach, producing a general widening. If the groins are continuous along a great length of the beach, the result may be beneficial, although there is always the problem of the beach that is downcurrent from the place where the groins have ceased. These beaches are certain to lose a moderate amount of sand, but the short jetties allow much of the sand to keep moving.

Sand Supply for Beaches

In some places quarrying of shell beaches by bulldozers has gone on for years with few interruptions, and yet the beach has continued to be almost exactly the same size. In such cases nearshore organisms must produce a continuous supply of shells for the waves to carry shoreward. The waves are working toward the maintenance of a profile of equilibrium, which man is constantly destroying by his quarrying activities. Thus, every time the bulldozers cut away the beach and steepen the foreshore, the waves introduce new material that will restore the natural slope. Of course, there is the question of what would happen to the sand if it were not being quarried. Excess sand brought in by the waves might be swept along the coast after the profile equilibrium was developed, so that beaches downcurrent would get a larger supply if man were not removing this local excess. Considerable study is certainly necessary before one can settle points of this sort.

There are many places where the supply of shells being produced by marine animals along the shore never comes in to the beaches. For example, coral reefs sometimes form continuous barriers along the shore, preventing the waves from carrying in the small shells or coral debris necessary for a sand supply. Perhaps intelligent dredging of channels through such reefs would allow access of the material to the beaches. Otherwise the shell debris will continue to be carried out into deeper water, where it is deposited or is gradually broken to pieces by the waves and by various organisms until it is too small to form a beach.

In a rather surprising way man has indirectly destroyed some beaches. Rivers are the most common source of sand for beaches because the rivers transport sand along their channels until it eventually enters the ocean. Therefore, any-

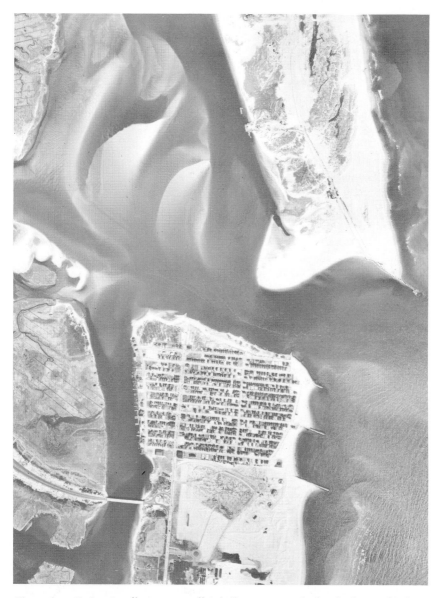

Figure 22. Groins (small concrete walls) built to protect the beach along a barrier at Ocean City, New Jersey. The buildout on the south side indicates a northerly current at this point. Shown inside the inlet are slightly submerged bars created by deposition from the tidal currents, which lose their power after entering the bay. A longshore bar is indicated along the outer coast.

thing that cuts off this supply of sand will tend to ruin the beaches. Dams, built for flood control or water supply along many of the rivers in dry areas such as southern California, where sand was formerly carried to the sea by the occasional heavy rains, have removed a large source of sand. The sand now collects in the lakes developed by the dams and therefore does not get to the coast. This is a matter that has to be given careful consideration, particularly in those resort areas that are dependent on their beaches for their sources of income.

Beach Characteristics

Coarse versus fine sand. You may have noticed that some beaches always have steep foreshores and others have gentle slopes. Along the east coast of Florida, for example, there is an alternation between steep foreshores, as at Miami or Fort Lauderdale, and the gently sloping foreshores, as at Daytona Beach or Fort Pierce. Generally when you walk along steep beaches, your feet sink deep into the coarse sand, whereas when you walk along gently sloping beaches, the fine sand is almost always firm under foot. Even automobiles can drive along the latter, as they do at Daytona. Many of the long beaches of Texas also are hard enough for automobile traffic, although the 130-mile Padre Island beach off southern Texas can be negotiated only by a jeep or truck because there are extensive shelly zones, where an ordinary car will get stuck.

The difference between these two types of beaches is easy to understand. The finer sand grains become readily interlocked after being deposited and thus develop a firm footing, whereas the coarse sand and shells pack loosely and are easily disturbed. Similarly, in the case of the coarse sands and shells, the water from waves sinks in rapidly, depositing the sediment carried in during the uprush until the slope has become sufficiently steep to allow the effective back-wash that will stop the growth. On the other hand, on fine sand beaches with smaller pores the water does not sink in as fast and hence can run back more readily, so that equilibrium will develop on a gentler slope. For similar reasons, the coarse sand beaches have a landward-sloping berm, whereas the fine sand beaches have berms that are nearly horizontal.

Except locally, the relative coarseness of sand is not related to exposure to waves. It is generally due to the source material. If coarse sand is provided locally by torrents from the mountains, by weathering of coarse-grained rocks, or by waves attacking coral reefs, a coarse-sand beach will develop. If fine material is provided by rivers or waves, the beach will be of fine sand. At Hanalei Bay in Kauai, northernmost of the Hawaiian group, the beach varies from coarse to fine sand along its length, depending on the source of supply. Near the mouth of the Hanalei River the sand is fine, but where the source is the coral reefs, the sand is coarse. The slopes and firmness underfoot vary accordingly.

Beach cusps. Beaches have many characteristics not commonly observed by bathers. Beach *cusps* are one of the most striking of these features (Fig. 23). These consist of a series of points or short ridges of sand or gravel protruding seaward with small embayments in between. These cusps are usually quite evenly spaced. They are close together in areas that have small choppy waves and far apart in areas where there are large pounding breakers. With relatively fine sand they vary in proximity from as little as a foot or less between cusps in beaches of small lakes to several hundred feet along beaches where ten-to-twenty-foot breakers come in from the open sea. Furthermore, the size varies from time to time. During storms, cusps many develop that are a hundred feet or more from crest to crest. These are often left at a higher level of the beach than normal waves because of the increase of the sea level during storms. During succeeding periods of quiet water small, more closely spaced cusps may develop at a lower level. The spacing is also dependent on grain size, in general the coarser the sand or gravel, the closer the spacing.

Figure 23. Beach cusps at San Simeon, California, formed in a gently sloping fine sand beach. The horizontal beach berm is shown to the left.

Cusps are the result of water that piles over the top of the berm and runs back along restricted lanes, producing depressions and leaving ridges in between. The latter are shaped into points by the waves. Cusps are far more common when the tide range is small, as during the neap tides of the first and third quarters of the moon. They tend to be destroyed during the large spring tides at the times of full and new moon. As a result, a trip by car along a beach may be more comfortable during the period of a full or new moon.

Troughs and bars. Similarly, the submerged *troughs* and *bars* that develop at or near low-tide level are far more pronounced during periods of neap tide because a wide-ranging tide tends to shift them between different levels of the beach rather than develop them more completely at one particular level. These troughs and bars are usually exposed at low tide (Fig. 24) along the

Figure 24. Bar and trough partially exposed at mid-tide. Photograph taken near St. Petersburg, west coast of Florida.

areas that have small tidal range, as for example on the Gulf Coast or in the Mediterranean, where the tides have a normal range of not more than one or two feet.

Ripples, rills, and domes. Low tides often reveal a series of *ripples* extending along fine sand beaches. These are largely the result of the backwash or runoff of the receding waves, which sets up a turbulent motion. Similar action takes place to a certain extent on coarse sand, but the water sinks into the beach more readily and thus does not develop as much turbulence. Furthermore, the large sand grains settle back into place more readily. Ripples may also be caused by current motion in the troughs, and these troughs are partly exposed at low tide (Fig. 25).

Rill marks are another phenomenon observed at low tide (Fig. 26). These are small channels caused by seaward drainage of the water that has been stored in the beach at high tide. At low tide this produces miniature valley

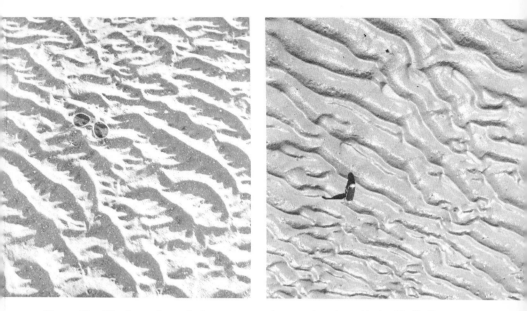

Figure 25. Ripple marks made by currents and exposed at low tide in Cholla Bay at the upper end of the Gulf of California, Sonora, Mexico. Photograph by Tad Nichols.

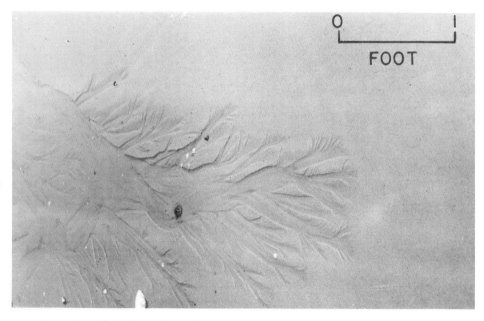

Figure 26. Rill marks made by the water running out of the beach sand exposed by low tide. Note the streamlike pattern of the rills. Photograph by Tad Nichols.

systems on the lower beach. Sometimes when water is returning seaward from a wave, its motion may be interfered with by a shell, a pebble, or perhaps the projecting antennae of little animals buried in the sand, such as the small sand crab (Emerita). The antennae are pushed up above the beach sand in order to collect small organisms. The deflection of the back-flowing water by these antennae tends to produce a diamond pattern (Fig. 27). Shells or pebbles protruding out of the beach sand have a similar effect.

Figure 27. Backwash marks made by projecting shells, pebbles, or perhaps by the antennae of small sand crabs (Emerita), which deflect the backrush of the waves.

In some places the rush of waves over an area entraps and confines air. This air expands and lifts the sand, making miniature *domes* an inch or more high and from a few inches to a foot in diameter. Digging into these domes you find they have a hollow center and may collapse as the air escapes. Waves often cut off the tops of the domes and develop black and white rings because of the layered nature of the underlying sand that has been domed upward.

Beach stratification. The layering in beach sands can be seen on the side of almost every hole dug into the sand (Fig. 28). This beach characteristic is in part the result of the variation in the types of material that are deposited under different conditions. For example, if the waves are particularly small, the beach may develop a cover consisting largely of shiny mica. This makes it glisten with a yellowish light that raises in the uninitiated false hopes of gold. This mica is easily carried away by large waves and may be entirely removed by an ensuing period of heavy surf. It may also be covered up by another layer of different material, such as quartz. Small shells may be brought in during quiet periods, forming another type of layer on the beach.

Figure 28. Layering in beach sand near San Francisco exposed on the side of a cut. The disturbance in the stratification is probably produced by entrapment of air.

At other times the beach sand, which has been building up for weeks or even months, may be winnowed by storm waves, which concentrate the heavier iron-bearing minerals, such as magnetite, previously scattered through the sand. These heavy minerals are commonly black, and thus a black layer may form. If you cut through to the bedrock that underlies many beaches, you may find black layers at the very bottom of the sand, especially in the upper beach near sea walls or cliffs. These black layers represent the concentration of storm periods when the beach has been largely stripped away. Some of the black sands contain rare and valuable minerals, such as tungsten, scheelite, and wolframite. Beach layering under the berm is apt to be horizontal, but below the sloping foreshore the underlying layers have an inclination that is roughly parallel to the foreshore slope.

Along many of the beaches in tropical areas there are sandstone rocks with a layering that has almost exactly the same slope as that of the adjacent beaches. This is known as *beachrock* and is the result of percolating waters that have cemented together the sand grains. To the uninitiated, beachrock is sometimes mistaken for elevated coral reefs, which may also exist along the shores in tropical areas. An examination of the beachrock, however, will show the absence of individual corals and of the complicated structures that characterize the coral reefs (see Chap. X).

The above is only a partial description of the numerous phenomena that characterize a beach. Careful inspection will probably show you many other interesting things. Almost all of these special features change or disappear from time to time, mostly in response to changes from rough to calm conditions or in the direction of wave approach. One can certainly enrich the enjoyment of a beach vacation by observing and trying to understand the cause of what you see.

CHAPTER IV

THE CONTINENTAL SHELVES
THAT SURROUND THE LANDS

To geologists, the shallow platforms that fringe the continents are of interest because they represent the environment in which a large portion of the sedimentary rocks were deposited in ancient seas. To oil companies, the continental shelves have taken on very considerable importance in recent years because they are proving to be one of the last possible sources of the vast petroleum supplies that are required in our civilization. To mariners, the continental shelf has long been an important factor because it is on the shelf that soundings are used widely as an aid to navigation. To most fishermen, the shelf is necessary for their livelihood because it is from these shallow areas that most commercial fish are obtained.

Defining the Shelf

In 1946, by a presidential decree, the United States took possession of the mineral rights to the *continental shelves* adjacent to our territory. Other nations will certainly follow our lead. This being the case, it is of considerable importance to define the continental shelves as clearly as possible. Otherwise, nations will be claiming rights in areas that are completely incomparable. In virtually annexing the shelf, we defined it as the shallow-water area extending out to a depth of 100 fathoms. This, however, is purely arbitrary and does not take into account that the word *shelf* refers to a flat area with a steep termination. There are such flat areas with rather abrupt edges around most of the continents, and it is only rarely that they terminate at or even close to 100 fathoms.

In 1953, the International Committee on the Nomenclature of Ocean Bottom Features defined the shelf as the "zone around the continents, extending from low-water line to the depth at which there is a marked increase of slope to greater depth." This outer slope is called the *continental slope*. In most cases this definition is adequate, but in some localities there are two or more breaks in slope, so that legally it is rather important to distinguish between these breaks. A recent UNESCO conference, in which I participated, resulted in the suggestion that where there are two breaks, the most marked break be used, provided it lies at depths of less than 600 meters or 300 fathoms, according to which unit is employed by the nation.

It was formerly supposed that most continental shelves were flat, rather evenly sloping areas, but this has not proven to be the case except locally.

Elsewhere, the shelf includes areas as hilly as southern New England, which no one would refer to as flat. Despite its irregularities, however, the continental shelf does have a shelf aspect, as can be seen from a series of profiles that include both the flat and the uneven varieties (Fig. 29).

Types of Shelf Sediment

Prior to describing the numerous varieties of continental shelves, some attention will be given to the common types of sediment that are found on the shelf. Of these, sand, which consists of grains with diameters between 2 millimeters[1] and 1/16 millimeter, is the most common. The sands include three principal types. The first type, *terrigenous sand,* is made up of minerals derived from the breaking up of rocks on land by weathering. The small fragments are carried to the sea largely by streams. The most common constituent of terrigenous sand is quartz, but many other minerals such as feldspar, mica, hornblende, and augite are included. *Calcarenite sand,* the second type, is made up of shells (or shell fragments), foraminifera, coral debris, and other organisms that consist of calcium carbonate. The third type, *authigenic sand,* is formed in place either by direct precipitation from the sea water, as in the case of the concentric onion-like oölites, which are a form of calcium carbonate, or by replacement of some other sediment like the greenish mineral, glauconite, or the brown phosphorite.

These three types of sand are referred to respectively as *physical* for the terrigenous because they are introduced by physical processes; *organic* for the calcarenites because they consist of the remnants of organisms; and *chemical* for the authigenic sands because they are due primarily to precipitation following some chemical reaction in the sea water. The three varieties are often found mixed together in the same sediment. In such cases the sand is usually named according to which of the three predominates. It is important to distinguish the three types because they help interpret the origin of the sediment. For example, a purely terrigenous sand generally indicates that there is a large source of minerals from the land and rapid deposition. Authigenic sands suggest slow deposition because it takes time for the authigenic minerals to form, and they would be masked by rapid deposition. The predominance of calcarenites is indicative either of abundance of organisms living in the area and providing the sediment or of an absence of a good source of terrigenous minerals.

The second most abundant type of sediment on the shelves can be referred to as *mud* (often called *lutite* by geologists). This sediment has a predominance of grains with diameters less than 1/16 millimeter. Technically these are referred to as *silts* if the diameters are between 1/16 and 1/256 millimeter, and

---

[1] 25 millimeters = one inch.

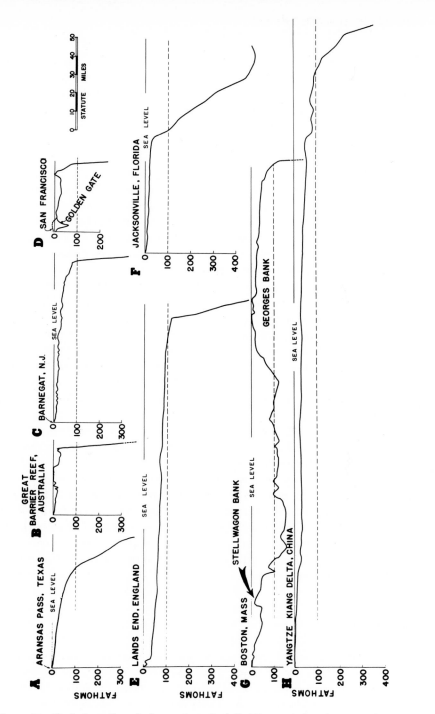

Figure 29. Typical profiles of the continental shelf. The vertical scale is greatly exaggerated. This serves to show the variation between a smooth graded slope (Sec. A) and the highly irregular shelf off a glaciated area (Sec. G).

as *clays* if the diameters are less than 1/256 millimeter. Many sands are easily distinguished from muds because their grains can be clearly seen. In borderline cases the muds can sometimes be distinguished from the sands by drying the sediment and noting whether it adheres. The fine sands will fall apart, whereas the muds will usually remain somewhat cohesive, so that they have to be dis-aggregated to determine grain size by soaking in various solutions such as sodium hexametaphosphate.

A third type of sediment, much less common on the shelf than sand or mud, is *gravel,* consisting of material coarser than 2 millimeters in diameter. Gravel is further divided into *granules* if less than 4 millimeters in diameter, *pebbles* if from 4 to 64 millimeters, *cobbles* if from 64 to 256 millimeters, and *boulders* if over 256 millimeters. Other relatively rare shelf sediments are volcanic fragments, such as bombs, ash, or pumice (a porous volcanic glass); seaweed or other types of marine plants such as kelp, which may be growing on the bottom or may exist as mats of the dead plants; oysters, which grow in large colonies, mostly in brackish bays; and sponges, which abound locally on the open shelf in tropical areas.

On many parts of the continental shelf our only source of knowledge of the sediments comes from chart notations made by the surveyors from small samples taken during their operations. Often these chart notations are inaccurate. For example, there is rarely any differentiation by surveyors between the types of sand, although the calcarenites may be referred to as coral sands or as shell sands. Muds are referred to as mud-and-sand or sand-and-mud if the surveyor observed grains of sand in the soft muddy sediments. The use of these terms varies a great deal among surveyors, as we have found from comparing our analyses of sediments with the chart symbols.

Extensive areas on the continental shelves are free of a sediment mantle. These areas may have either a rock bottom, which consists of any of the major rocks found on the continents, that is, sedimentary, igneous, and metamorphic, or a bottom that is covered with reef corals, coralline algae, or other organisms that consolidate as they grow, so that they develop a hard bottom. Some areas where there are cobbles or boulders that were impossible for the surveyors to distinguish from ledge rock are referred to on charts as rock bottom.

Bottom Contours

Although charts show depths largely as soundings, there is an increasing use of contour lines (called depth curves on charts) to bring out the features of the bottom. If it were not for the distaste for innovations among navigators, contour charts of the ocean bottom would have long ago superseded soundings in all but the shoalest areas. Contours consist of lines connecting points with the same elevation, that is, height above or below sea level. In land maps, contours indicate that the areas inside the encircling contour are higher than those outside. If, however, there are basin depressions, the contours are hachured (area marked "DEPRESSION" in Fig. 30), and the land inside the contour is then by definition lower than that directly outside.[2]

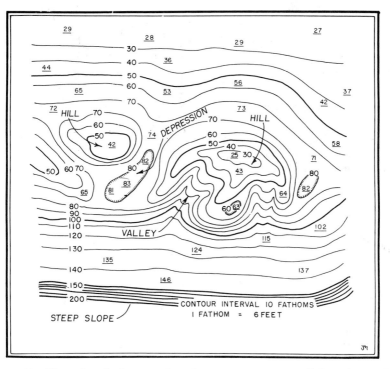

Figure 30. Illustrating the interpretation of a contour map as applied to the sea floor. The underlined numbers represent soundings.

In Figure 30 a hill is indicated by a series of somewhat concentric circles, a valley by a series of V-shaped indentations in the side of a hill, and a basin by its hachured contours. The relative steepness of a slope can be told by the closeness of spacing of the contours. If the slope is vertical, the contours are shown as a bundle without any space between them. In judging slope steepness, it is important to refer to the contour interval, that is, the vertical difference between two contours. This interval is set by the cartographer according to the map scale, the detail of available information, and the amount of the relief. The contour map of the Grand Canyon has a large contour interval,

[2] In marine navigation charts, hachuring is not used, although it is used by geologists for submarine contouring.

despite the abundance of information available, because of the precipitous slopes.

Continental Shelf Types

Although there are still many poorly surveyed shelves around the world, enough information is available, particularly around the coasts of the United States, so that a number of very distinct shelf types have been discovered. Some of these are distinguished by their topography and others by their characteristic sediment patterns, but in most cases both the relief and the character of bottom sediment are considered. The shelf types are quite closely related to the geology or the relief of the bordering lands. The types to follow will be described with little attempt to give origin, which is considered largely in Chapter V.

Shelves bordering glaciated land masses. The first type to be considered is in many ways the most interesting. If all continental shelves were like those bordering the glaciated areas, it is somewhat questionable whether the term *shelf* would have been applied. In these areas, the sea floor is strikingly irregular, and the term *shelf* is applicable only in the broader sense of the word and as contrasted with the continental slope.

Along most glaciated coasts, deep bays extend far into the land either as fiords like those of Norway and British Columbia or as deep troughs like the St. Lawrence or the straits of Juan de Fuca. Out beyond these embayments the deep troughs continue, often crossing the entire shelf, but usually having shoaler depths on the outer shelf than in toward the land. Since these deeps commonly have depths of more than a hundred fathoms and since they often extend in close to the shore, they form ample justification for our definition, which does not limit the continental shelves to depths of less than a hundred fathoms. The deeps contain many basins, which is also true of the adjacent land masses, although on land the basins are generally filled with water, and hence represent lakes and are not contoured on the topographic sheets. The shelf basins usually have muddy sediments, but the mud always contains a sprinkling of gravel and sand.

The shelves off glaciated areas have an abundance of hills. On the inner shelves most of the hills have rock bottom or are covered with boulders and other types of gravel. On the outer shelves there are many banks extending along the shelf and rising close to sea level, so that they form a marked contrast to the deep troughs and basins that are more common on the inner shelves. Most of these banks are covered with sand, but gravel of all sizes is found also. The banks include some of the best fishing grounds in the world. For example, Georges Bank off the New England coast (Fig. 31) supplies the bulk of the fish for the large cities of the northeastern United States, whereas

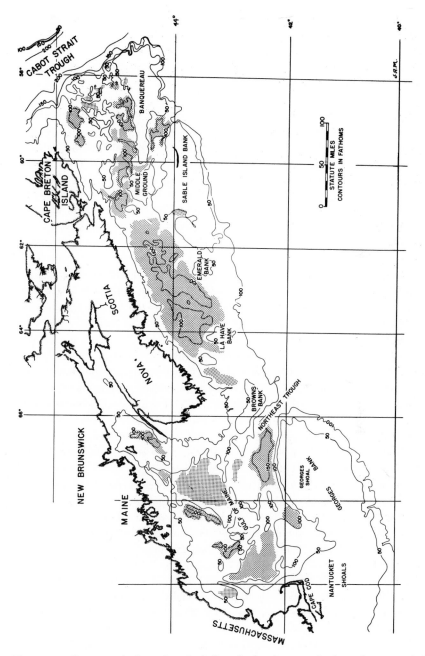

Figure 31. Contours of the series of shallow banks that extend along the outer shelf north of Cape Cod. Shaded zones are basins.

Banquereau off Nova Scotia and the Grand Banks of Newfoundland form a supply not only for Canada and parts of the United States but also much of western Europe. The elevation above the surrounding shelf increases current action, which introduces food nutrients into the area, causing the plankton to thrive and hence develop food for the fish.

If you were to take a trip out from Boston and run to the east on a boat equipped with a recording fathometer, the sounding trace would show you the various relief features characteristic of a shelf bordering glaciated lands (Fig. 29, Sec. G). You would first cross a relatively shallow basin in Massachusetts Bay and then rise over a sand ridge, Stellwagon Bank. Then the trace would indicate that you were crossing a series of deep basins and troughs before rising over Georges Bank, where you might be in some danger of getting caught in the breakers that form on top of some of the shoaler portions. Beyond this crest you would observe a gradual deepening and smoothing of the sea floor until you reached the break in slope at the shelf edge at about sixty fathoms.

Shelves with elongate sand banks and depressions. The shelves off glaciated areas form a class by themselves. The other types, considered as a whole, are comparatively flat, so that the term *shelf* seems more appropriate for them than it does for those off glaciated areas. The change in shelf character from rough to relatively smooth takes place quite abruptly off the east coast just south of Cape Cod and southwest of the shoals of Georges Bank. Highly detailed charts of the zone to the south show a low relief with a series of somewhat curving ridges (Fig. 29, Sec. C; Fig. 32). Between the ridges are shallow depressions, which in general are parallel to the ridges. This type of topography is somewhat like that found on the bordering coastal plain of New Jersey. The ridges and depressions are comparable to the sand ridges of the coastal plain with their intervening marshes, except that the low areas and salt flats are flatter on the coastal plain, presumably because they have received more sediment.

The sediments off the mid-Atlantic states have been studied as part of a large project of the U. S. Geological Survey in cooperation with Woods Hole Oceanographic Institution. This work, under the direction of K. O. Emery, has shown that the inner shelf is largely sand-covered with scattered patches of gravel, and the outer shelf has fine-grained and more uniform sand. Near the head of Hudson Canyon (Fig. 32), authigenic sediments are found along the outer edge of the shelf with glauconite and foraminifera shells partly filled with pyrite. Mammoth bones and shallow-water shells, dated by carbon-14 as thousands of years old, have been found by Woods Hole geologists on parts of this shelf.

The elongate banks extending across the shelf south of New York appear to be reasonably stable. Other elongate sand banks are subject to constant fluctuations and therefore form a serious menace to navigation, since the charts frequently become obsolete. Examples of shifting sand bars are provided by

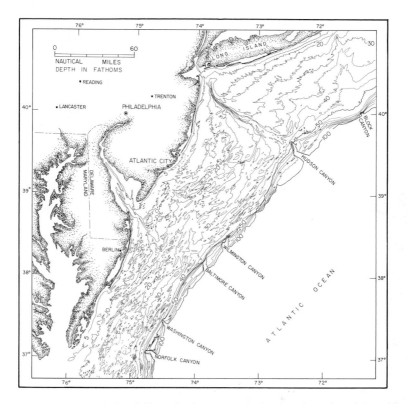

Figure 32. Contours of the shelf south of New York showing the series of low ridges that are comparable to the sand islands along the present shore. Shallow valleys crossing the shelf are also indicated, as are the heads of submarine canyons penetrating the shelf margin. Redrawn from A. C. Veatch and P. A. Smith, Geological Society of America, Special Papers, No. 7, Sept. 30, 1939 (Plate 1).

the Nantucket shoals. These are shifted by severe storms, as I learned when skirting Nantucket after a small hurricane. The boat I was navigating ran ashore in an area where the chart showed plenty of water. At least, grounding on shifting shoals is not as serious as landing on rocks, so that after kedging with an anchor and shifting weight from side to side we worked our way off into deep water.

Another locality of shallow shifting sand banks lies in the North Sea just north of the English Channel (Fig. 33). These banks run roughly parallel to both coasts, and one group extends up the Thames estuary. The area is so important to navigation that it is very well charted, and the shifting of the shoals is watched constantly. Great changes occur during storms like that of February, 1953, when the lowlands of Holland and England were so badly flooded.

Shelves with strong current action. There are several types of shelves associated with strong current action. One is found along the east coast of Florida

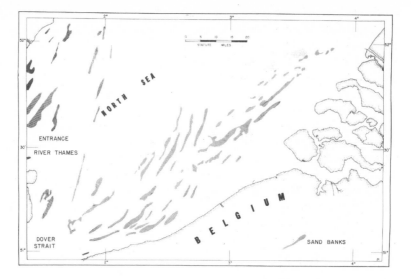

Figure 33. Sand banks that form shifting shoals along the southern portion of the North Sea.

where the Gulf Stream, flowing north from the Straits of Florida, swings close to the shore. From a width of about 60 miles at Jacksonville, near the Georgia line (Fig. 29, Sec. F), the shelf decreases to a width of about a mile off Palm Beach and there is virtually no shelf farther south. Even off Jacksonville the shelf is much shallower than average, having shelf-edge depths of thirty fathoms as compared to sixty or seventy off the mid-Atlantic states and New England. As the shelf narrows to the south, it also becomes shoaler, becoming only a few fathoms deep off Palm Beach. Despite this narrowing and shoaling of the shelf on the east coast of Florida, the shelf along the entire west coast of the same state is about 150 miles wide and has depths that are almost as great as those off the northeastern states.

The Florida shelf where swept by the Gulf Stream is covered with coarse sands, gravels, and water-worn shells. Many of the sands are authigenic (chemically precipitated in place), consisting of phosphatized particles along with glauconite and oölites, all of which indicate precipitation from sea water rather than transportation of the particles to the area. The worn shells appear to have formed in coastal waters and are now found in somewhat greater depths than those in which the organisms lived originally. Apparently not much, if any, deposition is now occurring on the shelf.

Also associated with strong currents are the deep holes that occur where the currents are concentrated. These holes are especially pronounced at the mouths of bays, as in the Golden Gate, where the water is 384 feet deep (Fig. 29, Sec. D), and at the Bungo Strait entrance to the Sea of Japan, where there is a hole 2,580 feet deep. The strong currents moving through this strait and the other entrances to this inland sea made it almost impossible for the Japanese to guard against enemy submarines during World War II. Nets and mines were difficult to maintain because of the combination of the strong currents and the deep water.

Shelves bordering large deltas. Several types of shelf are found along the borders of large deltas. The Mississippi River has extended its bird-foot delta (Fig. 43) in very recent years across the entire width of the shelf, so that at South Pass a slope continues directly down to the deep Gulf of Mexico. There are shelves, however, on either side and these contain terraces at depths of 138, 216, and 300 feet. Outside the Indus Delta there is a much wider shelf with a great terrace at 300 feet almost as large as the delta itself. Off the Niger (Fig. 34) and the Nile deltas the shelf bends seaward outside the deltas so

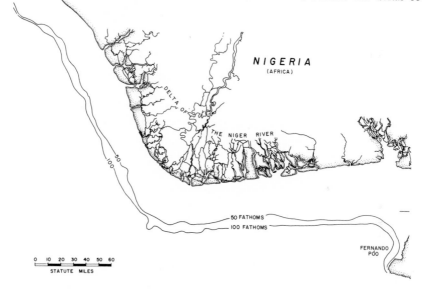

Figure 34. Illustrating how the continental shelf margin bends out parallel to the front of the Niger Delta.

that the edge in each case runs approximately parallel to the curvature of the delta.

The sediments around the margins of the large deltas are almost entirely mud, much of it with a high clay content. These fine sediments contain an abundance of wood fragments carried out from the rivers. You can understand this if you cruise down the lower reaches of a large river and observe how the banks are lined with snagged tree trunks. The sand fraction of the sediments off rivers is usually very high in mica, which is more easily transported by river currents than rounder sand.

Off most large rivers the zone of mud is bordered by sand. This is found off the Orinoco, the Amazon, the Yukon, and off most of the large rivers of Asia, especially eastern Asia. In this arrangement we find the most complete refutation of the theory that the grain size of sediment decreases seaward. Figure 35

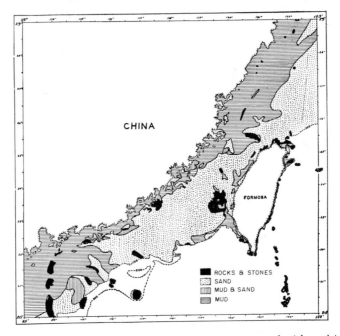

Figure 35. Showing how the outer shelf off eastern Asia is covered with sand in contrast to the muddy sediment near shore.

shows the relation of sand and mud to distance from shore on the shelf off the East China Sea.

Smooth shelves with sediment size diminishing seaward. The type of shelf that formerly was supposed to be typical, that is, one in which there is a relatively smooth slope out from shore and on which the sediments become finer with distance from shore, is actually so rare that it is only in the last few years that we have found a type example, and even in that case the shelf has a few hills on its outer side that are covered with relatively coarse sediments. The area where this supposedly normal type was found is off Aransas Pass along the central Texas coast (Fig. 29, Sec. A). Near shore the sediments are sandy like those of the barrier islands, only somewhat finer. They change rather abruptly to muds, and the clay content increases fairly constantly across the shelf. On the outer shelf, foraminifera make up almost all of the small sand content, and these foraminifera are mostly of the globular type called *Globigerina,* which drift near the surface while alive and after death sink to the bottom. Nearer shore the foraminifera in the sediments are dominantly bottom dwellers.

Possibly there are other nicely graded shelves. Some marine charts show sand near shore and mud on the outside, and the soundings indicate fairly even slopes. Unfortunately we lack samples from these places, and the soundings are rather widely spaced.

A shelf with salt-dome hills. In 1935 I was informed by my friend Captain Frank Borden of the United States Coast and Geodetic Survey that he was charting some hills on the outer continental shelf west of the Mississippi Delta. My curiosity was aroused, and I joined the party on the next trip out from New Orleans. After a long hot trip down the Mississippi, we entered the Gulf of Mexico, following along the outer edge of the muddy current emerging from Southwest Pass. We headed south-southwest and in seventeen miles found that the flatness of the shelf was interrupted by a hill. The weather was calm and there was scarcely a ripple on the water. Under these ideal conditions we completed the survey of the hill in less than a day and found that it had a shape much rounder than most shelf hills (Fig. 36). It rose from a shallow surrounding moat with a depth of about 370 feet to a 200-foot summit. It was not easy to sample the summit of the hill because it proved to have a hard surface, but after several tries we succeeded in bringing up a mass of coralline

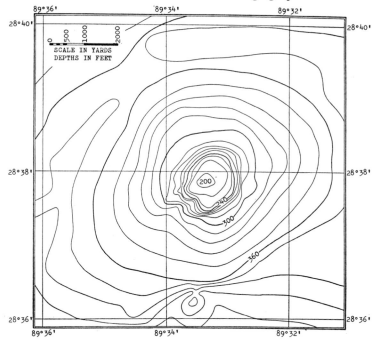

Figure 36. An oval hill rising from the outer shelf southwest of the Mississippi Delta and probably produced by a great intrusion of salt from underneath.

algae, which showed that despite the close proximity of the mouth of a large muddy river, lime-secreting organisms were able to establish a hold on the bottom.

Visiting the United States Coast and Geodetic Survey office in Washington somewhat later, I obtained the detailed unpublished soundings of a whole series of somewhat similar banks along the shelf edge west of the Mississippi Delta. These other hills are not quite as round, but many are encircled by moats. Knowing of the salt domes discovered along the coast of the Gulf of Mexico inside this area, I suggested, perhaps rather rashly, that the outer shelf hills were *salt domes*[3] and that they stood up above the general shelf because of the absence of erosion and solution, which have succeeded in removing most of the relief in the salt domes on land in this area.[4] About ten years later Henry Stetson, operating from the Woods Hole Oceanographic Institution vessel *Atlantis,* visited several of these hills and dredged from them numerous round algal masses along with some reef corals. He made the alternative suggestion that the hills represent coral reefs, or what is referred to by geologists as *bioherms* (hills that owe their growth to some type of calcareous organism).

The question of whether the shelf hills are salt domes or bioherms soon developed more than academic interest. The Texas and Louisiana salt domes are not only an economic source of sulphur and salt, but even more important, they are flanked by tilted sediments, which have yielded enormous supplies of oil.[5] Accordingly, as soon as offshore drilling became feasible, the oil companies began to investigate the shelf hills by geophysical methods. Most of their results are still company secrets, but gradually it has become known that at least some of the domes have a gravity relationship that indicates that salt plugs are present. The salt is heavier than the adjacent water-saturated sediment, so that higher gravity is indicated.

In our Scripps Institution work on the Texas continental shelf we found still other evidence favoring salt-dome origin for the hills. We were cruising along the outer shelf seventy miles off Galveston on a calm day when the water was so clear that it was possible to look down on one of these banks, some sixty feet below, and to distinguish clearly rock ridges extending along the long axis of this elliptical bank. Divers tried to descend onto the bank with scuba gear, but they were surrounded by mean-looking barracuda who may have been only curious but looked for all the world as if they were hungry. As a result, the scientists decided to content themselves with dredging. The rock brought up in their dredges proved to have Miocene fossils (deposited about

---

[3] These are oval masses of salt, often a mile or more in diameter, which rise with vertical walls for thousands of feet through the surrounding sediments. They come from deep salt beds and have been pushed up because of the plasticity of the salt.

[4] Exceptions are High Island on Bolivar Peninsula, Texas, and Avery Island, Louisiana.

[5] The rising salt bends up the sediments. The oil, being lighter than water, migrates up to where the beds are sealed off by salt.

25,000,000 years ago), and yet the Miocene layer on the Texas shelf is known to have a cover of sediments amounting to about 10,000 feet. This means that this outer-shelf dome has been pushed up at least 10,000 feet to get the Miocene material to the surface. Our oil-company sponsors of the project that led to this discovery were understandably skeptical of our findings until we gave their paleontologists samples of the foraminifera with which we had established the ages.

Elsewhere there are no definite evidences of salt-dome hills on the continental shelves, although in the enclosed Persian Gulf numerous submerged hills are believed to have that origin. It will be surprising if more domes are not discovered as exploration of the shelves continues.

Shelves rimmed with rocky banks and rocky islands. Many of the narrow continental shelves, particularly those off mountainous coasts, are characterized by extensive areas of rocky bottom. The rock ledges are especially well developed along the outer shelves. This is illustrated by a map of the shelf off the San Diego, California, area (Fig. 37). Here Coronado Bank forms the outer shelf margin rising above a submarine valley of low relief that lies in toward the land. The rocky nature of the bank was made evident to Scripps Institution scientists by the large number of rock dredges that we lost there during wartime investigations when old worn-out wire had to be used. To the

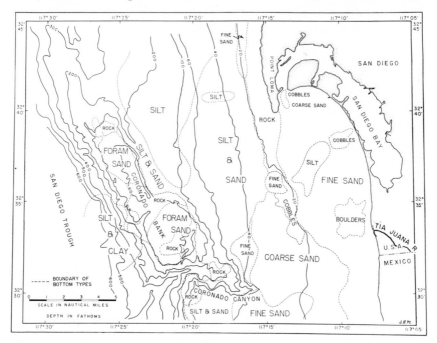

Figure 37. The somewhat haphazard distribution of sediment types on the shelf off San Diego, California. (Foram. stands for foraminifera.)

south of the bank, beyond a canyon that indents the shelf, the rocky floor continues and rises to form the rugged Coronados Islands belonging to Mexico. The rocks of the outer shelf bank and islands proved to have been deposited during the Miocene and Pliocene, millions of years in the past. The strata in the islands are considerably deformed, and the same appears to be true of the bank. A photograph of the bank surface is shown in Figure 38.

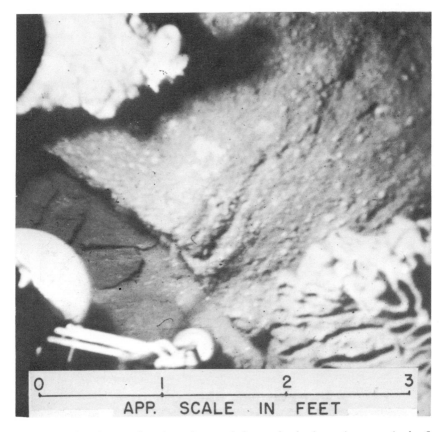

Figure 38. A curious rock surface photographed at a depth of 380 feet on a bank off San Diego. In the upper left is seen an apparently overhanging rock ledge with a small coral. Object in lower left is a portion of the photographic equipment.

One of the interesting things about the Coronados Islands is that, instead of being surrounded by shallow wave-cut benches, their flanks descend abruptly in many places to depths of fifty or a hundred feet. While diving onto one of these submarine cliffs with scuba gear, I was impressed by the extensive caves and the overhanging rock masses. The huge kelp plants that grow up from the sides form a wonderful shelter for the bright orange-colored garibaldis and many other fish that weave in and out of the great fronds and add to the majestic scene in this swaying underwater forest.

The shelf off San Diego inside the rocky rim is not devoid of interest. Like so many other shelf areas, the grain size of the sediment does not grade from coarse near shore to fine farther out. Instead, moving seaward from the Tia Juana River (south portion of Figure 37), one encounters first a fine sand like that of the beach and then a boulder patch where there are water-worn rounded cobbles and boulders with only a little rather shelly sand in the interstices. This patch is several miles wide and despite its flat surface seems never to have been covered with the fine sediment that is carried out of the Tia Juana River during floods. Beyond the boulders there is a continuation of the fine sand, then a patch of cobbles, which in this case contains considerable fine sand and even silt. Next is a coarse sand that has a brown color in contrast to the gray finer sand near shore. The brown color is due to oxidation and is rather common in stream deposits. Farther out the sediment becomes silty, although it retains a considerable amount of sand. In the bottom of the little valley that lies shoreward of Coronado Bank, small cobbles were dredged in one haul, although fine sediment is obtained here ordinarily. On the rocky Coronado Bank there is a thin discontinuous cover of calcarenite sand. Foraminifera are the most common constituent, and many of these are filled with glauconite, which suggests that deposition has been very slow, since glauconite is a slow-forming authigenic mineral.

The Santa Monica Bay shelf off the Los Angeles area and the shelf off San Francisco (Fig. 29, Sec. D) have many of the same features described for the San Diego area. Fishery charts of the shelf off west Portugal suggest that here also is another example. In fact the San Diego shelf appears to be typical of narrow shelves in general, both in its rocky bottom and in its patchwork of sediment.

Shelves with shallow discontinuous valley systems. In a few places there are valleys that are incised only slightly below the general level of the shelf, in contrast to the deep troughs of glaciated areas. The best-known example is the Sunda shelf, lying between Borneo on one side and Java, Sumatra, and the Malay Peninsula on the other (Fig. 39). The soundings are not abundant enough to establish the pattern for sure, so there is some uncertainty in the interpretations given by the Dutch geologists G. A. F. Molengraff and Philip Kuenen that the valleys consist of a series of branching valleys with the dendritic (leaf vein) pattern characteristic of many land river systems. Locally, near the islands, there are channels related to tidal scour but connected with the dendritic valley system, although out on the shelf the tides are thought to be very weak and hence cannot explain the valleys.

Another better-sounded but less continuous series of shelf valleys is found off the mid-Atlantic states (Fig. 32). The valley that heads directly off New York Harbor is the best developed. It virtually crosses the entire shelf, dying

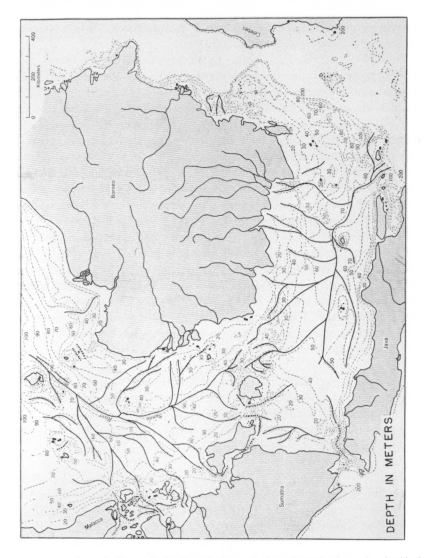

Figure 39. Chart of the Sunda Shelf with drowned river channels shown by the black lines. Note their relation to the island rivers. From Philip Kuenen, Gröningen University.

out near the outer edge in a feature that has been compared to a delta. Along the valley course there are various elongated basins but none of a depth nearly as great as those in the troughs off the glaciated coasts. The channels coming out of Delaware and Chesapeake bays terminate a few miles outside the bay, but in Chesapeake Bay the valley can be traced for a hundred miles, as far up as Baltimore. Borings show that these valleys are partly filled with sediment, so that Chesapeake Bay is only 150 feet deep, but the rock floor is as deep locally as 300 feet below sea level.

The English Channel (Fig. 40 and Fig. 29, Sec. E) has another valley, Hurd Deep, between Devonshire and the Cherbourg Peninsula, where the water attains a depth of ninety-four fathoms within ten miles of the island of Alderney. Here the tide can be presumed to have played an important role, because the English Channel has tides that are among the highest in the world. The currents flow here with such force that the sea develops a particularly nasty chop, causing much *mal de mer,* as the writer knows from sad experience. Rocks are exposed widely on the floor of the English Channel. They have been investigated by the French and English. They are of special interest because of the possible tunnel under the channel that has long been considered as a means of connecting the two countries.

Shelves shoaled by coral growth. In many tropical areas, moundlike or elongate reefs rise above the general level of the shelf. In some places these reefs rise from shelves far too deep to allow the reef-forming corals to live. These reefs will be described in Chapter X, but no discussion of shelf types would be complete without mentioning them. The best-known example is the Great Barrier Reef off northeastern Australia (Fig. 29, Sec. B), where the reefs cover hundreds of square miles and convert a wide area into a danger ground for navigation except during daylight hours, when the shoals are easily seen from the masthead in this fogless area.

In one sense the Australian barrier reef shelf compares with shelves off glaciated areas. The deeper water is near the coast, and the dangerous reefs are mostly beyond, but the inner channel is much shoaler and smoother, and the outer reefs reach the surface in many places in contrast to the glacial area banks.

Scuba Diving and Future Shelf Studies

Since the average depth of the shelves, approximately 30 fathoms, is within the limit of scuba dives made by experts, one can confidently expect many important geological discoveries in the shelf investigations in the near future.[6] Already groups of scuba divers working for oil companies have explored many areas off southern California. Among their discoveries have been anticlinal

---

[6] *SeaLab,* the U.S. Navy project at 200-foot depth off La Jolla, California, and *ConShelf,* the Jacques Cousteau project at 300 feet off the French Riviera, demonstrated the capacity for work on the continental shelf by men living in specially constructed houses and breathing various gas mixtures to prevent the bends from developing under the great pressures of those depths. They could move out in the water with scuba and for rest and meals return to the houses that were anchored on the bottom. Both operations continued for several weeks. It is now anticipated that depths of approximately 1,000 feet may become practical for this work in the future.

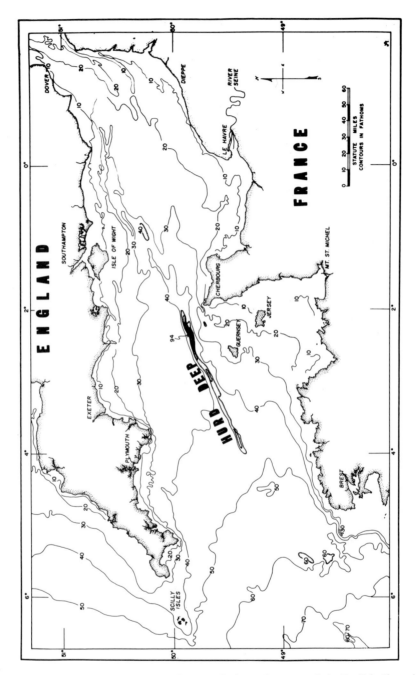

Figure 40. The discontinuous valley that extends down the center of the English Channel.

domes, which are interesting because they are often underlain by oil-bearing formations. The underwater swimmers have also seen asphalt deposits, which are slowly emerging and building mounds on the sea floor. In localities where there is a thin sediment cover, the scuba divers often have jetted their way through to the underlying rock formations, using hoses lowered to them from vessels on the surface.

Exploration of the inner banks off Texas has been accomplished by the aqualung-equipped geologists of Magnolia Oil Company. They have made interesting discoveries of the faunas that could be obtained only by great effort by using the ordinary sampling methods. The school of French scuba divers associated with Jacques Cousteau, who founded this rapidly growing technique, has added an important chapter to archaeology in dives on the shelf along the Mediterranean coast.[7] Their well-known discoveries of Roman galleons and of their amphora jugs have been a great stimulus to other explorers using this method. Our Scripps Institution scuba divers operating around La Jolla have discovered a great abundance of metates either left on the shelf by the Indians when the sea level was lower or, less likely, carried out from the land in canoes.

Shelf Statistics

A number of generalizations may be made concerning the continental shelves. The average width appears to be close to forty-two miles, the average depth of shelf edge is seventy-two fathoms, the average depth of the flattest portions of the shelf is thirty-five fathoms, and the average shelf slope is 0°07', or ten feet per mile. These figures may prove of some importance in the problems of international law, particularly in view of some of the claims now being made on territorial limits by nations that are bordered by very narrow shelves.

The shelves are deepest off glaciated areas, although there are a few deep shelves elsewhere, for example, off Southwest Africa. The shelves are shoalest in areas with extensive coral growth and along unglaciated Siberia where it faces the Arctic Ocean. The shelves are widest in the Arctic and along the north and west sides of the Pacific from the Bering Sea to Australia. The shelves are narrowest off young mountain ranges where numerous earthquakes indicate that intensive faulting is still taking place.

The shelves are covered with a diversity of sediment types, reminiscent of a patchwork quilt in most areas. No rule of decreasing or increasing grade size in crossing the shelf can be made, but there are few exceptions to the curious generalizations that wide shelves off large rivers usually have fine sediment near shore and coarse sediment farther out.

[7] Jacques-Yves Cousteau, "Fish Men Discover a 2,200-year-old Greek Ship," *Nat. Geog. Magazine,* CV (1954), 1-36.

CHAPTER V

ORIGIN OF CONTINENTAL SHELVES

Earlier Explanations

Geologists for a long time explained the continental shelves as a product of present-day conditions. The inner shelves were referred to as *wave-cut terraces* because of the inroads of wave action onto the lands, whereas the outer shelves were depicted as constructional platforms, or *wave-built terraces,* caused by the deposition of the material eroded by the waves or introduced by rivers (Fig. 41). The outer depths of the shelf were said to represent *wave base,* that is, the depth below which wave action was supposed to be ineffective.

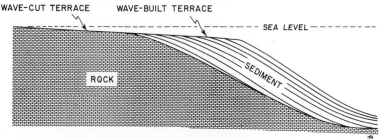

Figure 41. Illustrating the theoretical concept of a continental shelf, consisting of a wave-cut terrace inside and a wave-built terrace outside.

As soon as information became available showing the true condition of the continental shelf, this simple explanation ran into many difficulties. The discovery of extensive areas of rock bottom on the outer shelves provided a contradiction to the wave-built-terrace idea. In some areas where rock does not occur on the outer shelves, the old idea is invalid for other reasons. First, rock is often found just below the margin of the shelf, either on the slopes or on the walls of submarine canyons cut into the shelf, and hence there could be no appreciable wave-built terrace. Second, the depths of the outer shelves are not related to the size of existing waves as one would expect if these depths were due to wave base. Thus, the shelf edges on the sides of continents exposed to great storm waves, instead of being deeper, average a little shoaler than those on the more protected sides. Third, the sediments are so variable in size distribution on the outer shelf that it is inconceivable that they could represent an outgrowing sediment embankment with an edge at the depth where waves lose their power to stir the bottom. Fourth, the topography of much of the outer shelf is too irregular to be indicative of a wave-built terrace. In spite of these objections, recent evidence from continuous reflection profiling (see p. 172)

83

has indicated that many of the wider shelves have grown outward by accretion, which is similar to the wave-built terrace. Probably much of this growth has been at sea levels relatively lower than those now existing.

Sea-Level Lowering During the Ice Age

The most important factor in explaining the present condition of the continental shelves appears to be the lowering of sea levels that occurred during various stages of the ice age (Pleistocene period). At this time great continental glaciers covered large portions of North America and somewhat smaller areas of Europe and South America (Fig. 42). This ice must have come from the ocean by evaporation and precipitation as snow and hence resulted in lowered sea level. The volume of the ice has been estimated by various scientists,

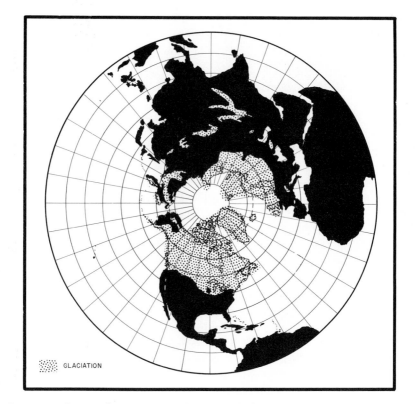

GLACIATION

Figure 42. Showing the vast extent of continental glaciers in the Northern Hemisphere during the ice age.

and when translated into sea water, these volumes suggest that the level of the ocean was lowered from about 250 to 500 feet.[1]

Unfortunately, all estimates of ice volume have been handicapped by a lack of knowledge of the maximum thicknesses of the ice sheets and an uncertainty as to the contemporaneity of the greatest ice advances. Many of the estimates have been too low because the scientists who developed them did not know that the great glaciers had extended onto large areas of the continental shelf, which, as will be explained below, is indicated by the submarine topography. Furthermore, the earlier belief that various highlands within the glacial margins were never covered by ice has now been largely dispelled. Such areas as the Gaspé Peninsula in Quebec, northern Labrador, and the highest portion of the Norwegian Peninsula were thought at first to have been ice free because of lack of glacial deposits and because of their flora, which was thought to be endemic and a relic of preglacial times. Detailed studies by geologists, however, have shown that these supposedly ice-free areas have glacial deposits and glacial striations. It is believed by some investigators that the absence of easily recognized glaciation at these places may be the result of the intense frost action. Alternatively, these areas may not have been glaciated during the last major glacial stage but were ice covered in an earlier stage so that much of the evidence has disappeared, and floras have had time to develop their endemic characteristics during the interval between some early glaciation and the present.

Methods of Detecting Glacial Sea Levels

Actually, the best means of estimating the volume of continental glaciers may come from a study of the continental shelf. If we assume that the maximum development of the glaciers lasted for an appreciable period, such as a thousand years, indications of the sea level at this time should be found in the underwater topography. Waves working on soft material would cut cliffs and terraces at a rapid rate, and at least the inner limits of terraces produced in this way should be close to the sea level of the time when they were cut. If the current Antarctic and Greenland ice caps should melt (thus raising the sea level), the present-day cliff and terrace contact would be submerged in numerous places all around the world, leaving evidence of the sea level as it stood prior to the melting. Following such a submergence, however, deposition might occur at the base of the cliffs, so that the terrace surfaces would gradually disappear. In this way the rise in sea level that followed the maximum glaciation submerged the terraces and cliffs of that low level, but some of the terraces are now buried.

---

[1] The latest information coming from new soundings and continuous reflection profiles places these estimates at about 500 feet for the next to the last glacial stage.

The former German island Helgoland, an important wartime submarine base in the North Sea, has been largely cut away during historical times, leaving only a very small island surrounded by a large bank that is just below sea level. During the lowest stands of sea level, islands on the outer shelf may have been beveled in the same way as Helgoland. The banks thus produced should now lie just below the glacial-stage sea levels.

Deltas represent another type of feature that could help establish the greatest lowering of sea level during the ice age. Deltas grow very rapidly at the mouths of large rivers, and most of this growth takes place at or very close to sea level. For example, in about a hundred years the Mississippi Delta added a strip of territory to Louisiana amounting to at least a hundred square miles (Fig. 43). All of this new land is within a foot or two of sea level. If the Mississippi should change its course to follow the shorter Atchafalaya channel to the east, the level of the present delta would sink a few feet due to compaction of the underlying muddy sediments. Sand deltas, on the other hand, would not undergo sinking from this cause. In either case, deltas built during glacial lowering might still be found at approximately the level of the maximum glaciation.

Because of the pounding of the breakers, the ability of waves to erode is certainly much greater in the surf zone than in the deeper areas farther from shore. Therefore, wherever the outer shelf has been cut by wave action, the level of the outer edge may be only a few feet below the lowest glacial sea level. Hence, the marginal depths of the shelf may be helpful in determining the lowest levels of sea during the ice age.

With the use of the foregoing lines of evidence, one might think it would be a relatively easy task to find where the sea level stood during the glacial maximum. Unfortunately there are many complications. First, many of the old eroded surfaces apparently have been covered with sediment. An example of such burial was found by jetting into the sediment on the shelf off Santa Barbara, California (Fig. 44). Here an old terrace had been largely blanketed by the recently deposited sediment. Further difficulty comes from the general instability of the earth's crust. Wherever areas have recently sunk, terraces cut during lowered sea levels or deltas built during these times will now exist at a level below the stand of the sea that produced them. Similarly, in areas that have been elevated, the terraces will be at too high a level, even above present sea level.

Because of these earth movements, the best results should be obtained in areas where there has been little disturbance of the earth's crust to throw glacial-age terraces or deltas out of line. Actually, there appear to be no coasts that are entirely free of earth movements. If we refer to the coast of the United States, for example, it is commonly stated that the east and south coasts are stable, whereas the west coast is unstable. We know, however, that both the east coast and the coast of the Gulf of Mexico have been sinking for millions

Figure 43. Illustrating the large growth of the Mississippi Delta during the last century. Courtesy of U.S. Coast and Geodetic Survey.

of years because wells drilled along these coasts at many places penetrate thousands of feet of sediments, all of which, as far as we know from the fossils, were deposited in shallow seas or on land. These sediments extend back to at least the Cretaceous period (70 to 130 million years before the present). In order to bring these sediments to their present great depth, however, the rate of sinking needs only to have been of the order of one foot in 10,000 years, so that the sinking should not have had any appreciable effect on terraces de-

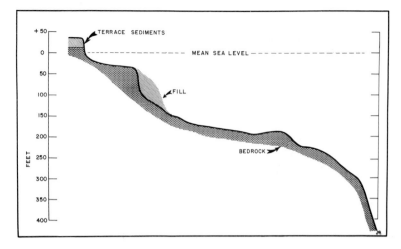

Figure 44. A wave-cut terrace drowned and partly buried by recent sediments. The existence of the terrace was established by jet borings made by a diver on the sea floor. Vertical exaggeration approximately times 10.

veloped during the last ice maximum (probably about 20,000 to 25,000 years before the present). Nor has the sinking been very significant in deepening the terraces of earlier ice stages, because the entire glacial period endured less than a million years.

In looking for reliable evidence of lowered sea level, one should be wary of those coasts where there are frequent earthquakes and where mountains are actively growing, for example, California, southern Alaska, Japan, the Philippines, and western South America. Actually, however, the marine terraces are not very different in level off some of these unstable areas from those found off the stable coasts, which suggests that movements have not been very great since the lowered sea levels of the ice age.

The average depth at the shelf margins, where the greatest change in slope takes place, is 72 fathoms (432 feet), and since we can assume that this is slightly below the maximum glacial sea-level lowering, we can think of that lowering as being approximately 400 feet. On the other hand, there are many terraces on the shelves with depths up to about 50 fathoms (300 feet). Below this depth, terraces appear to be quite rare, so that there is a possibility the 300-foot level may be better than 400. Alternatively, the 300-foot level may represent the sea stand of the last major glaciation, and the terraces at lower levels, formed by earlier glaciations, may have been largely blanketed with sediment, leaving only the break in slope at the shelf margin as an indication of the lowest level. This shelf break is maintained in many places because of strong currents that are encountered at a topographic discontinuity of this type.

In the near future it should be possible to track the rise of sea level at the end of the last glacial stage by dating with carbon-14[2] the shallow-water sediments left when the sea rose. Preliminary determinations indicate that the sea rose rapidly from about minus 260 feet up to minus 20 feet between 17,000 and 6,000 years before the present. After that time either the sea continued to rise slowly up to the present level, or it stopped rising some 5,000 years ago, or it stood some 10 feet above its present level during a time when the climate was warmer than the persent (4,000 or 5,000 years ago) and then lowered to its present level during historical times. My own opinion is that the evidence is more favorable to the idea that there has been a very slow rise continuing to the present.[3]

Cause of the Irregular Distribution of Sediment Size

If the sea had stood at the present level for a long time, the finer sediments would have covered the outer shelf, at least in places that lacked strong bottom currents. It seems clear also that in general the sediment would have become graded outward across the shelf from coarse near shore to fine outside. The failure to find this gradation can be attributed to the same changing sea levels that have developed the topography of the shelf.

When the sea level was lower, coarse shallow-water or land deposits were laid down in areas where now the water is too deep for these materials to be introduced. As definite evidence of this hypothesis, a few carbon-14 age determinations have shown that the shelly sands on the outer shelf off Galveston, Texas, were deposited either just before the sea level rose or during the rise. In many areas the sediments carried by rivers during glacial stages were coarser than those carried by present-day rivers because of greater volume of water in the rivers supplied by the melting glaciers. These coarse sediments were deposited on the exposed shelves.

The melting of the glaciers, beginning about 17,000 to 20,000 years before the present, caused the drowning or embayment of river valleys along most of the coasts of the world. The result of this drowning was to cause rivers to deposit most of their sediments in long-armed bays (like Chesapeake Bay) and hence to deliver very little material to the open shelf. For this reason little sediment has been deposited on top of the coarse shelf sediments formed during the last glacial stage. Even off some of the large delta-building rivers these coarse sediments remain uncovered on the outer shelf. Here, however, sediment appears to be available, but in many places currents are too strong to allow

[2] A technique used by radiologists for dating shells and plant remains in the age range of up to about 40,000 years old.

[3] Carbon-14 dates up to 1966 have provided extensive evidence of this slow rise in sea level during the past 6,000 years.

its deposition. Examples of these conditions are found on the shelf off such large rivers as the Amazon and the Orinoco, where the North Equatorial Current is very strong and undoubtedly extends to the bottom along the outer shelf. Off the great rivers of eastern Asia (Fig. 35) the Kuroshio (Japanese Current) is flowing at a high speed along the outer shelf and hence prevents mud deposition over a broad strip. On the other hand, around the Mississippi, and especially in a zone extending west of the delta, sediments are now being deposited on the outer shelf, and hence we find a postglacial cover over the coarser material that formed during glacial stages.

Present-day barrier islands, like those at Atlantic City, New Jersey (Fig. 32), Jones Beach and Coney Island off New York, and Galveston Island of Texas, have their counterparts on the shelf. These were developed during low sea levels and submerged by the rising sea. In areas where there has been little deposition the drowned islands are easily recognized. They occur now as the elongated hills on the continental shelf off the east coast (Fig. 32). After the outermost islands formed, the sea rose, but in rising to its present level there were many halts. During each of these halts new islands formed and were in turn submerged, so that eventually the entire width of the shelf became covered with sand ridges and intervening troughs.

Off the Louisiana and Texas coasts west of the Mississippi Delta other drowned islands have been discovered. These are less continuous, and some of them are thought to be related to salt domes. Borings by the Magnolia Oil Company, however, have shown that some of the questionable hills are simply sand deposits formed on top of the shelf muds, just as sand beaches along the present shore have covered old mud deposits. It is by no means impossible that scuba investigations will find artifacts of Indians who may have inhabited these islands during the sea-level rise.

The Shelf as Modified by Glacial Erosion and Deposition

It can scarcely be a coincidence that almost all of the deep troughs and basins of the continental shelves are found adjacent to glaciated land masses. The present shore line was no barrier to ice movement during glacial stages, partly because the sea level was lower and partly because glaciers were quite capable of advancing out into the ocean, as they do now along most of the perimenter of Antarctica. The initial advance of the ice could have been in the form of a floating tongue, but as the ice thickened it would have rested on the bottom. Once having contacted the bottom, the ice could wear down the former sea floor as easily as it did the land.

An examination of the topography in formerly glaciated land areas shows great irregularity. Lakes with depths of hundreds of feet are common all over such glaciated areas as Canada and northern Europe. In the United States, the Great Lakes, which are among the deepest as well as largest lakes in the world, all lie in glaciated territory. Mountain ranges that were formerly glaciated also have a myriad of deep lakes. Most geologists have come to believe that these deep lake basins are the result of glacial gouging.

It is a curious thing that many geologists who accept without hesitation the idea of glacial erosion on the lands, when confronted with similar basins and troughs on the sea floor, have turned to earth movements, principally faulting, as an explanation. Thus J. W. Gregory, a Scottish geologist who became famous as the explorer and interpreter of the Rift Valleys of Africa as fault valleys, suggested that the Norwegian fiords also were due to faulting. If he had investigated other glaciated coasts and found how similar the fiords in all of these areas are to those of Norway, he might have hesitated in making such an interpretation, particularly since he could not have found any typical fiords in unglaciated areas.

Douglas Johnson of Columbia University, famous for his books on shore processes, examined the charts off New England and concluded that the Bay of Fundy represented an old fault valley. Like Gregory, he failed to notice that similar features occur along virtually all glaciated shelves. While Johnson was familiar with glacial erosion forms, it apparently had not occurred to him that the glaciers could work beyond the continental confines. More recently Olaf Holtedahl, a Norwegian geology professor, formerly of the University of Oslo, and his son Hans Holtedahl of Bergen University, have been interpreting many of the shelf features off Norway as due to faulting. This may be partly correct, particularly since many of the escarpments extend parallel to the land, but again it seems strange that similar features are found off most other glaciated coasts and they have not been found elsewhere on the shelves, even in areas where, unlike Norway, recent faults and fault valleys are very much in evidence on the adjacent lands.

The cable breaks that accompanied the Grand Banks earthquake in 1929 were approximately in line with a great trough that comes out of Cabot Strait (Fig. 9). Shortly after the earthquake Arthur Keith, a geologist, and E. A. Hodgson, a seismologist, formerly of the Dominion Observatory in Ottawa, suggested independently that these breaks had occurred along faults that they believed had been previously responsible for the Cabot Strait Trough, and hence they suggested that the trough was a fault valley. Now, however, these breaks seem to be explained more acceptably as having been caused by landslides or turbidity currents.

About the time of the Grand Banks earthquake I had become impressed with the similarity of the Cabot Strait Trough to other glacial troughs, so I thought I would look for evidence of glaciation in Cabot Strait. Sailing north from Boston in a small yacht, after a few days we approached the Coast Guard Station at the southeast end of St. Paul Island in the middle of Cabot Strait. It was after dark, so we tried to get assistance in entering the cove. A somewhat inebriated Canadian came out in a skiff and piloted us in past roaring breakers. Next morning we found we were almost surrounded by jagged rocks. We landed, however, and set off down the length of the four-mile island to the northeast. Near the north end of the island we discovered glacial striations pointing down the Cabot Strait Trough, as I had hoped. Next morning we set sail for Port-aux-Basques, Newfoundland. There again we found striations to show that the ice had come out of the gulf through the strait. From this and other evidence I am left with the conviction that the Cabot Strait Trough is just another feature caused by glacial erosion.

Glacial deposition also has been important in developing the shelf topography in the glaciated areas. The well-known fishing banks (Fig. 31) have been formed primarily by ice deposition. Most of these banks extend parallel to the coast and probably have much the same origin as the glacial moraines that are formed at the margins of continental glaciers either by long continued dumping of glacial debris at the ice margin when the ice is melting as fast as it moves forward or by deposits built up by the streams emerging from the ice under the same conditions. The topography differs somewhat from the rolling hills of glacial moraines on the continents, but this is not surprising, since tidal currents have modified the moraines of the continental shelf, leaving shoals that are elongated in the direction of the major flow. The powerful currents on Georges Bank off New England have given the shoals a northwest-southeast trend.

In 1931, I accompanied a United States Coast and Geodetic Survey party during the charting of Georges Bank. We were anchored one night on a clean sand bottom that I had sampled. During the night a short but heavy blow came up, lasting for a few hours. Next day, when the anchor was raised, I collected a ball of clayey sediment from the flukes. This material proved to be full of sand and gravel fragments resembling typical glacial till found on land. The probable interpretation is that the surface sand is a concentrate left after the mud was winnowed out of the till by currents, but that under the sand a true till is still present. Evidence from another locality was obtained in recent borings made prior to building the two radar towers that were located on Georges Bank. In these borings, sand was encountered with only a little interbedded clay. Therefore, glacial till does not exist at these places. According to scientists at Woods Hole Oceanographic Institution, the borings yielded material similar to deposits on Long Island that were formed by streams emerging from the front of the continental glaciers. If streams also existed on Georges Bank, this is further evidence of lowered sea level during the times when the great glaciers extended out over portions of the continental shelf.

Shelves of the Preglacial World

The foregoing discussion has emphasized the importance of the glacial period in the development of the continental shelves and of their various characteristics. An interesting question, which is not easy to answer, is whether or not there would be a continental shelf if there had been no glacial period. We can be certain that none of the deep terraces and drowned deltas of the outer shelf could have formed without glacial lowering of the sea level. In preglacial times, however, there may have been millions of years during which a shelf was cut at one constant sea level by the waves.[4] Furthermore, the sinking of deltas could have added other large increments. Possibly also alluvial plains may have subsided, forming shelves over extensive areas. Whenever and wherever this subsidence of the relatively flat lowlands was more rapid than the upbuilding of deltas, shelves would have formed.

Shelves that developed during a still-standing sea level would certainly differ from those of the present day. There would be extensive erosion benches very close to the sea level of the time, but the benches would not be as deep as the outer shelves are today. The submerged plains and deltas would presumably have a cover of fine sediment, except near shore, and there would be little of the heterogeneity of shelf deposits now found.

Shelves would certainly be missing in a great many areas, whereas they are now practically universal. Most of the large rivers of the world would no doubt have built deltas completely across the shelves in their vicinity, whereas only the Mississippi has built such a cover since the last sea-level rise. On the other hand, smaller rivers and particularly rivers entering seas with strong coastal currents might not have succeeded in covering the shelf.

Without a glacial period, world commerce would have been more difficult, because there would have been few harbors of the embayment type, whereas these are now found indenting about two-thirds of the coasts of the world. The continental shelves would have been only a small fraction of what they are now. Shoal wave-cut terraces would have formed a widespread impediment to shipping. The present extensive shelf areas, where soundings reveal to the navigator the approach to the lands long before the depths become dangerous, would have been rare. It is indeed a fortunate thing that we had a glacial period.

Unfinished Business

The explanations of the continental shelves that have been given here are largely explanations of the present topography and sediments of the shelves. The nature of the underlying structure of the shelf constitutes another problem, as does the explanation for the abrupt edge of the shelf where it contacts the continental slope. Similarly, understanding the reason for the narrow shelf off northern Florida and the disappearance of the shelf off Miami requires some knowledge of the continental slopes. These features can be best discussed after considering the nature of the continental slopes in the following chapter.

---

[4] This would have been well above the present sea level because the ice locked up in the Antarctic and Greenland ice caps probably represents almost 200 feet of sea water. Thicknesses of ice are determined by setting off dynamite charges and timing the echoes received from the contact between the ice and the underlying rock.

CHAPTER VI

THE WORLD'S GREATEST SLOPES

If there were no ocean and we could look up at the earth from the moon, the most impressive relief features that would hit our eyes would be the continental slopes that extend down from the continents to the deep-ocean floor. If you have seen the mighty wall of the Himalayas from Darjeeling, you may wonder how the oceans could have anything as impressive, but actually there are many places along the margin of the Pacific where the slopes have a greater vertical relief. When combined with the adjacent land slopes of the Andes, the escarpment along the west coast of South America has a height of 42,000 feet, almost twice that of the south side of the Himalayas. Furthermore, great marine slopes extend all the way around the continents, whereas the slopes that border plateaus and mountain ranges on land are much less continuous.

It might be thought that despite their great vertical range the continental slopes would be far less impressive than the mountain slopes, when seen from the moon, because they would be so much smoother. Actually the differences between land and sea slopes in this respect are not very great. Probably the sea slopes have fewer minor irregularities, but they compare quite favorably in their major valleys and mountain peaks.

Defining the Continental Slope

Where shelves are present, there is no great difficulty in finding the inner or landward margin of the continental slope, since this boundary represents also the outer margin of the continental shelf, or in other words the starting point of the toboggan slope that continues down toward the oceanic abysses. More difficulty arises in finding the outer boundary, which separates the continental slope from the deep-ocean floor. In some places, notably around the Pacific, the slope shows little if any decrease until the greatest depths are attained. Elsewhere, however, the steeper portion of the slope is terminated by a very gentle apron-like slope that extends for scores or even hundreds of miles into the deep-ocean basins (Fig. 1). This outer slope is referred to as the continental rise or locally as a *deep-sea fan*. It is not only much less inclined than the inner slope, but it is also much smoother topographically. Accordingly it seems wise to confine the term *continental slope* to the steeper portions and to consider the apron as a part of the deep-sea floor.

Most continental slopes, although irregular, are quite continuous between the shelf and either the continental rise or the deep-sea floor (Fig. 45). In some places, however, there is an intermediate step existing either as a terrace (Fig. 45, Sec. H). or as a series of basins and ranges (Fig. 45, Sec. F). The basins are at shoaler depths than the deep-sea floor but at greater depths than the basins of the continental shelves, whereas the ranges or mountains rise to a variety of intermediate depths or actually come above the surface to form islands. These interruptions in continental slope continuity are called *continental borderlands*. Objections have been raised to this name, which I proposed in 1941, but objectors have so far failed to suggest any good alternative so the name was adopted by the International Committee on the Nomenclature of Ocean Bottom Features. Admittedly the use of the word *continent* and the suffix *land* for an area so far from the shore line could be confusing. Similarly our appropriation of the word *American* to apply to citizens of the United States is disturbing to our southern neighbors. In both cases, however, usage serves as at least a partial justification.

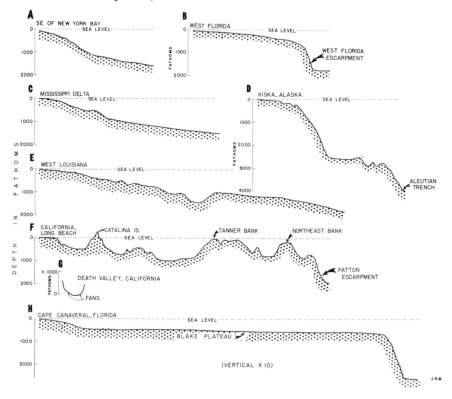

Figure 45. Comparative profiles of the continental slope, illustrating the various types. Note that a narrow shelf is shown vaguely inside some of the slopes.

Continental Slope Types

Just as the continental shelves are divisible into a number of rather distinctive types, so the continental slopes have several categories that are separated largely on the basis of topography. In choosing the examples it may seem unfortunate that they are almost all from slopes off the United States or United States territories. The reason for this apparent exclusiveness is that only the United States Coast and Geodetic Survey has made sufficiently detailed charts to allow description of the slopes from a two-dimensional point of view. Elsewhere only individual sounding lines extending down the slope are available, except in a few very restricted zones. Off the entire coast of the United States the slopes have been well surveyed out to a depth of 6,000 feet, and in many places these surveys have gone much deeper. Because of military needs the deep surveys have been extended to Alaska, with special emphasis on the Aleutian Island arc.

The canyoned slopes off the northeastern United States. If the sea level were lowered a mile, the people inhabiting the great populated belt that extends from New York to Washington would find that the most majestic scenery for their vacation trips was not in the Appalachians or the Adirondacks but on the continental slopes. Actually this scenery, which only a favored few can see in small doses from a deep-diving vehicle, begins eighty miles east of Atlantic City and about a hundred miles east of New York.

The continental slope off the northeastern United States is not precipitous. In fact it varies from 4° to 7°, but what makes this slope impressive is the continuous series of canyons cut into it (Fig. 32). At the deepest points below their rims, the principal slope canyons have a depth of from 2,000 to 4,000 feet. By way of contrast, the deepest canyons of the Appalachians are only about 2,000 feet deep. The entire continental slope is serrated down to depths of at least 6,000 feet, where the detailed surveys stop. Some fifteen of the canyons extend back into the continental shelf, notably the Hudson Canyon, which is virtually a continuation of the shallow shelf valley off New York.

The slopes off the northeastern United States are mostly covered with mud, but the mud in many places may be only skin deep, and rock outcrops on the canyon walls, particularly off New England. In one place rock was photographed on a part of the slope where there was no canyon (Fig. 46). The rocks of the canyons and slope contain fossils that make it possible to tell their geological age. They are all very old, Pliocene to Cretaceous.

The terraced slope south of Cape Hatteras and the Gulf Stream. A curious thing happens to the continental slope south of Cape Hatteras (Fig. 47). If you draw a line along the outer margin of the shelf north of the cape, you will see that it runs close to a steep escarpment, whereas the outer margin of the shelf south of the cape bends to the west, getting well out of line of the shelf edge farther north. Between the shelf edge and the outer escarpment to

trated to rocks about 50 million years old, which are found at far greater depths under the continental shelf and under the peninsula of Florida.

Much of the outer part of the plateau is covered with a mantle of manganese that forms a thick plate over the rock formations. In the inner plateau, there is a belt of phosphorite nodules. Both of these materials may have great economic value for mining in the future.

For some years, it has been known that the outer portion of Blake Plateau had some irregular hills. Recently, Woods Hole scientists took bottom photographs that explained these hills. They are ahermatypic[1] coral reefs of a type that, unlike the reef corals described in Chapter X, can live in deep and cold water.

Outside the Blake Plateau, there is an escarpment with slopes up to at least 30°. It is relatively straight but has a broad indentation to the south. On the escarpment, Lamont scientists found layers of rock as old as Early Cretaceous, approximately 100,000,000 years ago. This cliff is generally believed to represent a great fault, but it has been partly masked by sedimentation.

The origin of Blake Plateau is now tentatively explained as a slowly subsiding area like the continental shelf north of Cape Hatteras, but the Gulf Stream prevented building up of sediments on this plateau, whereas more active sedimentation maintained the shallow depths on the sinking northern shelf. The escarpment outside Blake Plateau is the southern continuation of the northern continental slope, but has not been subject to the erosion that produced the canyons to the north. These ideas may be subject to some change as the drillings from the *Joides* holes are further studied. Unlike the non-deposition on the Blake Plateau, the gentler inner slope has been built forward for many miles by sedimentation.

The escarpment outside the Blake Plateau is much steeper than the slope to the north. The soundings indicate inclinations of about 15°, but it is likely that there are truly precipitous slopes involved. At these great depths it is not easy to determine the exact angle of slopes. The escarpment is said to be cut by submarine canyons, but no maps have been published as yet.

The great "fault scarp" off western Florida. If geologists had been asked to predict the location of areas on the continental slope with fault scarps like the east face of the Sierra Nevada, they would never have guessed that the slope off western Florida had one of them. No earthquakes have been recorded from here, and the adjacent land is very stable. Yet the United States Coast and Geodetic Survey has discovered one of the steepest and straightest escarpments of the entire ocean floor extending for 500 miles along this slope (Fig. 48 and Fig. 45, Sec. B). Here the bottom drops a mile in a horizontal distance of less than two miles. There are few if any places on land where a plateau comparable to the Florida slope is bounded by such a steep escarpment.

[1] Those lacking algae living together with the corals in mutually cooperative life processes.

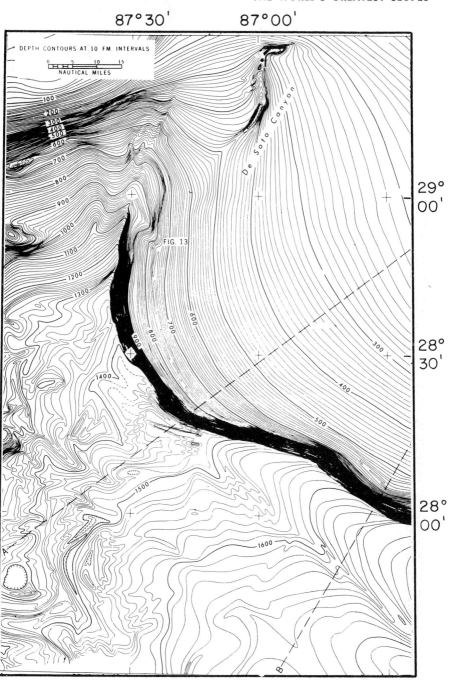

Figure 48. The unusually steep escarpment off western Florida, which is found in the lower portion of an otherwise gentle continental slope. Contours by G. F. Jordan, U.S. Coast and Geodetic Survey.

As far as we know, on the entire sea floor the west Florida scarp is exceeded in steepness only by the great fault scarp extending down to the 20,000-foot Bartlett Trough off Santiago, Cuba. In the latter there is an inclination of about 45°, or a drop of one mile[2] vertically in one mile. The escarpment off west Florida is fairly even in its inclination. Over most of the 500-mile length it is not cut by any appreciable valley, although to the south a series of valleys indent the deeper part of the escarpment. These valleys have floors that lie slightly below the level of the deep bottom of the Gulf of Mexico, which borders the escarpment.

The great submarine cliff west of Florida in no place contacts the shelf edge; rather it starts at depths of 3,000 to 4,800 feet. The continental slope shoreward of the cliff is quite gentle, some of it less than 1°, so that it is steep only compared to the adjacent continental shelf.

There has been little attempt to sample the escarpment. A few rocks were dredged from it by Woods Hole Oceanographic Institution. Also a photograph was obtained by Stetson in 6,600 feet of water that showed the rugged character of terrain that will be seen here when some day it is visited by a deep-diving vehicle.

The basin and hill slopes off Texas and Louisiana. North and west of the Florida escarpment the slope becomes as gentle and even as it is off Florida shoreward of the cliffed portion. Beginning west of the Mississippi River, however, it takes on a new complexion. There appear a series of hills and irregular but elongate basins (Fig. 49 and Fig. 45, Sec. E). If this surface were exposed, the basins would become lakes with depths of up to 1,500 feet and lengths of up to 30 miles. Since the elongation of the basins is mostly in the direction of the slope and since the basins are connected with slope valleys, it appears likely that the slope may once have been cut by valleys, but that something has blocked them.

Mrs. Betty Gealy, who studied and contoured the slope soundings in this area, concluded that the basins were the result of extensive landslides. This is a possibility, although the slopes are probably too gentle (about 1°) to allow the force of gravity to produce large basins like those found along escarpments. Furthermore, the submarine canyons that are common along most continental slopes are not known to be blocked in this way elsewhere, even where the continental slopes are much steeper. Another explanation is that the salt domes discovered on the outer shelf (Fig. 36) have pushed up portions of the slopes, and where these were thrust up into valleys, they formed a barricade.[3] In any

---

[2] The expedition of the U.S. Coast and Geodetic Ship *Pioneer* to the Indian Ocean, in 1964, led to the discovery of another equally steep escarpment off the east coast of Ceylon near Trincomalee Bay.

[3] This salt dome explanation has been confirmed in a number of cases by the continuous reflection profiling of Navy Electronics Laboratory, Lamont, Texas A. & M., and Scripps Institution scientists.

Figure 49. The basins and hills on the gentle slope off western Louisiana. Contours by Betty Gealy.

case the hills of the outer shelf quite clearly are found on the slope at considerable depth.

The slopes off Texas and Louisiana become much smoother toward the base, although in some places the basined area terminates in small fault scarps beyond which the smooth slope begins.

The borderland off southern California. Beyond a narrow and discontinuous shelf off southern California there is almost 150 miles of territory that is in most respects a counterpart of the basins and ranges of the adjacent land (Fig. 50 and Fig. 45, Sec. F). The mountains in this continental borderland rise as much as 8,000 feet above the level of bordering basins, and in the case of Santa Catalina, San Clemente, and Santa Cruz islands the mountains have summits that are about 2,000 feet above sea level. Some of the basins extend down to 9,000 feet below sea level. If the sea should be withdrawn and there were enough rainfall to keep these basins full,[4] they would contain lakes as deep as 4,000 feet, which is twice the depth of the deepest lake in the United States (Crater Lake in Oregon). The lakes would be from about ten to twenty miles across, and the longest lake (southeast of San Clemente Island) would be about eighty miles from end to end. If the basins did not have enough rain to fill them, they would be comparable to such desert basins as Death Valley. There appears, however, to be one difference: the basins of the borderland have flatter floors than the intermontane basins on land (Fig. 45, Sec. G). This difference is no doubt due to the great fans that are built out into land basins by alluvium washed from the mountains. Fans are also found on the sea floor, but in general the marine sediments appear to have been spread out more horizontally.

To the eye, there would be one striking difference between the California borderland and the basins and ranges on land. One would see far fewer canyons cutting the steep slopes of the basins on the sea floor than are found on land. The chief exceptions are off San Diego and north of San Nicolas Island. The steep escarpment beyond the outer limit of the borderland would look very different from the deeply dissected fault scarp on the eastern side of the Sierra Nevada, even though it has about the same angle of inclination (Fig. 51).

There seems to be little doubt that the escarpments of the borderland off southern California are due primarily to faulting, just as the fronts of the principal ranges on land are fault scarps, although modified by erosion. At the base of some of the seafloor scarps there are long depressions, which are sometimes found also at the base of land scarps. On land these ditches soon become filled by wash from the slopes, whereas those on the sea floor have more chance of being preserved. The rift valleys along the San Andreas Fault in southern California are supposed to be due to horizontal movement along this fault, which is thought to be accompanied by a pulling apart of the land surface. Similarly, a rift type of valley is found to the southeast of San Clemente Island (Fig. 50) where it looks as though Fortymile Bank might have been torn from the island by a large horizontal movement along a fault. In fact, it is possible that both Fortymile and Thirtymile banks were moved southeast as a block

---

[4] More than our present rainfall in southern California would be needed.

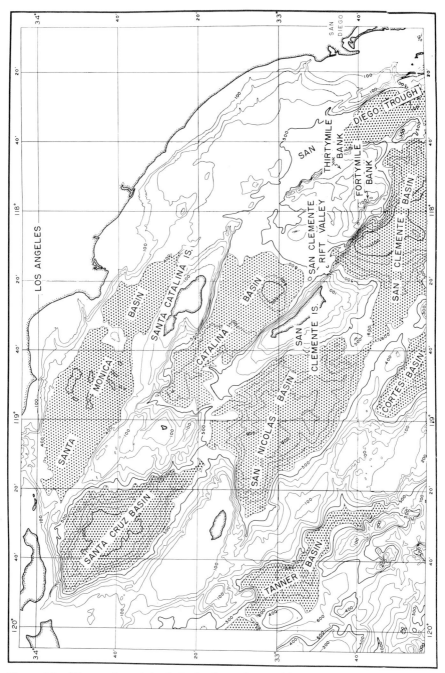

Figure 50. The continental borderland off southern California. Contour interval 100 fathoms. Shaded zones represent basins. Contours by Francis Shepard, Scripps Institution of Oceanography, and Kenneth O. Emery, Woods Hole Oceanographic Institution.

105

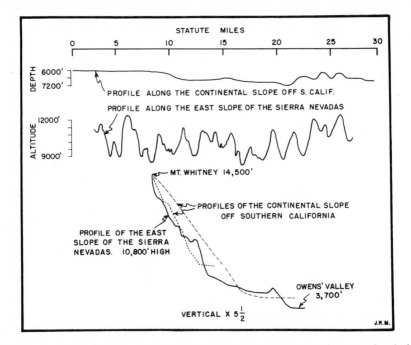

STATUTE MILES

PROFILE ALONG THE CONTINENTAL SLOPE OFF S. CALIF.

PROFILE ALONG THE EAST SLOPE OF THE SIERRA NEVADAS

MT. WHITNEY 14,500'

PROFILES OF THE CONTINENTAL SLOPE OFF SOUTHERN CALIFORNIA

PROFILE OF THE EAST SLOPE OF THE SIERRA NEVADAS. 10,800' HIGH

OWENS' VALLEY 3,700'

VERTICAL X 5½

Figure 51. Comparison between the slopes and the fault scarp on the east side of the Sierra Nevadas and the Patton Escarpment at the outer edge of the continental border-land off southern California.

from between Santa Catalina and San Clemente islands. This, however, is largely speculation based on the shapes of the banks and the escarpments and has not been verified as yet by the geological formations.

A westward-moving continental slope. After the great San Francisco earth-quake, geologists were much intrigued to find that the movement along the San Andreas Fault that caused the catastrophe had been in a horizontal direc-tion, so that roads and fences were offset as much as 20 feet with no vertical component. This fault has been traced almost the entire length of the state. From time to time other earthquakes have occurred along it, and each, as far as could be determined, has had the same type of horizontal movement with the west side of the fault moving relatively in a northward direction. Early studies of the San Andreas Fault led to the belief that over a long period there might have been many miles of total displacement of the same nature as that which occurred at San Francisco. For many years this idea of major shifting of the earth along this line was challenged by prominent California geologists, but more recently various careful studies seem to have left little doubt that the horizontal movements have been of a large order, so that entire mountain ranges have been displaced.

Except for the offset of features such as streams and roads, the effect of large horizontal movement is not shown on a flat land surface, whereas the displacement of a slope is clearly indicated. For this reason it had seemed to me probable that in northern California where the San Andreas Fault runs out to sea, there would be a major displacement of the continental slope. Just such an interpretation is suggested by a feature that is found near the westernmost point of the United States coast (Fig. 52). Here a contour map of the east-

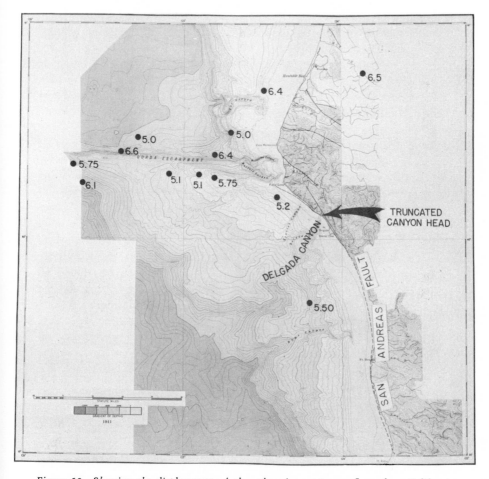

Figure 52. Showing the displacement of the submarine contours off northern California where the San Andreas Fault may bend seaward. According to this hypothesis, the Gorda Escarpment resulted from the seaward shift of the area to the south. The dots are earthquake epicenters and the accompanying numbers are magnitudes on the Richter scale, which measures the intensity of earthquakes according to their damage to structures. It will be noted that the large earthquakes (6 and over), are concentrated along this escarpment. Contours by Francis Shepard, Scripps Institution of Oceanography, and Kenneth O. Emery, Woods Hole Oceanographic Institution.

west Gorda Escarpment indicates a possible seaward shift of the slope on the south side amounting to about 40 miles for the outer part of the slope.[5]

When it was first suggested that this escarpment might be the continuation of the San Andreas Fault, the seismologists offered objections, saying that this was not the place where the earthquakes occurred off northern California. Recently, however, the compilation of the major earthquakes for the area by the California seismologists quite reverses this opinion, as can be seen in Figure 52. Furthermore, there is a line of gash cracks or small rift valleys along the continental shelf running diagonally from the Gorda Escarpment and resembling some of the rifts along the San Andreas Fault on land. Farther south, a submarine canyon seems to have been cut off by the fault and shoved along the coast until it abuts against a mountain wall rather than lining up with the valleys on land. All in all, it appears quite reasonable that the San Andreas does bend out here, although beyond the limits of the map it may bend back again to the north in line with the general trend of the fault. Earthquake epicenters on the sea floor to the north suggest this return.

The continuous reflection profiling by J. R. Curray and R. D. Nason of Scripps Institution has verified that the San Andreas Fault lies along the coast, as indicated in Figure 52, although the main fault may be several miles farther seaward than I had suggested.

Complex Aleutian slopes. Extensive surveys around the Aleutian Islands have revealed a slope topography that is distinctive in various aspects. On the south side of the islands the slopes lead down to the deep Aleutian Trench, where the water is more than 24,000 feet in depth. The slopes near the islands are cut by many valleys, but these valleys are not the river-canyon type found off both the northeast and parts of the west coasts of the United States. Instead they are trough-shaped, more like the basins off southern California, except that they slope outward quite continuously (Fig. 53). They have steep sides, suggestive of fault scarps, and rather broad floors. Furthermore, they run diagonal to the general slope, unlike river valleys. Also, the lower portions of the valleys have very low gradients and in places connect with a broad terrace at about 10,800 to 15,600 feet.

In 1966, the Coast and Geodetic Survey issued a series of contour charts showing the entire slope off the Aleutian Islands. The charts show many of the fault valleys south of the Aleutians, but an erosional type of valley is indicated in several places on the north side of the islands.

Beyond the terrace, on the south side, a steep escarpment leads down to the bottom of the Aleutian Trench (Fig. 45, Sec. D). As far as can be ascertained from the soundings, the slope of the outer escarpment does not exceed about 15°, therefore it is not so steep as the escarpment west of Florida. It also differs

---

[5] Victor Vaquier of Scripps Institution has reported recently that the magnetic anomalies indicate similar displacements on the deep sea floor off California, one with as much as 600 miles of movement (see page 156).

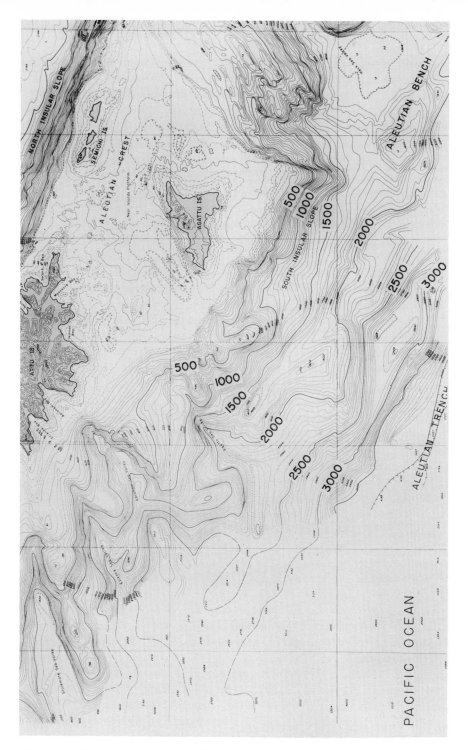

Figure 53. Illustrating the nature of the slope south of the Aleutian Islands. Most of the submarine valleys have a broad base and straight walls and are presumably the result of faulting. Contours in fathoms. From O. Gates and W. M. Gibson, Bulletin of the Geological Society of America, Vol. 67, Feb. 1956, No. 2 (Plate 1).

from the escarpment off Florida in being very active seismically. Many large earthquakes originate here, so there seems to be little doubt that it is a live fault. It is one of the greatest sources of tsunamis for the entire world. To the north of the islands another escarpment (North Insular Slope) has steeper slopes than those that lead down to the Aleutian Deep. Portions of this slope north of Attu Island have inclinations of about 25°, comparable to the escarpment west of Florida. This north Aleutian scarp extends from the shelf virtually to the deep-sea floor without terraces or other interruptions.

The increasing inclination of the lower slope south of the Aleutians is found also on various other continental slopes, notably off eastern Japan into the Japanese Trench and off southeast Australia. The continental slope off Japan also resembles that south of the Aleutians in having large trough-shaped valleys, but these lack basins.

Other slopes. Outside of United States territory most of the continental slopes for which there are sufficient soundings to judge their general topography are found bordering very narrow shelves. Several such slopes have been surveyed in the Mediterranean by the French. All of them are characterized by many submarine canyons, apparently of the dendritic river type found off the east coast of the United States. Those along the French Riviera, contoured by Professor Jacques Bourcart of the Sorbonne and his assistants, have amazingly intricate topography. Virtually the entire slope into the Mediterranean is as deeply eroded as are the sides of typical mountain ranges. Even the west side of the island of Corsica has a submerged slope like that found on the mountainous island above water (Fig. 65). Other examples of deeply dissected slopes are found along the east side of Formosa, along portions of eastern Korea, on the west and north sides of Luzon in the Philippines, and off Ceylon.

On the other hand, a few sounding lines run along the slopes have shown relatively smooth and even inclines. The really surprising thing in our present state of knowledge is that these even slopes are so rare, considering all of the millions of years of sedimentation that could have smoothed them.

Slope Statistics

In characterizing the continental slopes of the world, we can do little more than give some averages of inclinations shown by the soundings down to the 6,000-feet curve, which is as deep as most surveys have continued. The average slope from shelf edge to 6,000 feet is 4°07'. Most of the profiles obtained from the soundings indicate uneven slopes with depressions that, judging from the surveys off the United States, may be either valleys or basins. Slopes are more gentle off large rivers, especially off those with deltas. Here averages are 1°20', whereas off coasts that appear to be affected by faulting the slopes average 5°40'. Some of the steepest slopes in the world, however, are found off lands where no indication of recent faulting or of earth movements is found, for example, the great scarp off west Florida, the steep slope off the Brazilian Highlands, the slopes bordering various Bahama Islands, and the continental slope off southwest Australia and Ceylon. The average slopes of the Pacific are steeper than those of the Atlantic and Indian oceans. Respectively, these are 5°20', 3°05' and 2°55'. The Mediterranean slopes are intermediate with 3°34'.

The sediments on these slopes of the world, known largely from chart notations, consist of about 60 per cent mud and 25 per cent sand, whereas rock and gravel comprise about 10 per cent and the rest are labeled as shells or ooze. This is different from the shelves, where sand is more common than mud, although the percentages of rock and gravel are not dissimilar.

Origin of the Continental Slope

The term for continental slope in French is *talus continental*, indicating that the slopes are great talus piles built out beyond the shelf. Continuous reflection profiles have shown that this is partly justified, but the numerous rocky outer shelves and even rocky slopes indicate the danger of generalization. Beyond such deltas as the Niger (Fig. 34), the slope does appear to have been formed by a forward-building mass of sediment, and we know from comparison of old charts with new that the gentle slope off South Pass at the Mississippi Delta is being built forward at a rate of 200 feet a year. Many of the reflection profiles off the east coast of the United States and along the west coast of France show forward building of sedimentary layers conformable with an underlying slope. These depositional slopes have a low gradient, mostly less than 2° (see Fig. 45, Sec. C), which contrasts with more typical slope inclinations. Furthermore, the slopes of the world are cut by numerous escarpments, and in places steepen at depth rather than getting more gentle, as would be expected from a talus or sediment fan. The 10 per cent of rock and gravel reported from the slope is also contrary to expectations on the assumption that most of the slopes are depositional.

There are many reasons for believing that the continental slopes were at least originally formed by faulting, although many of them may have had no recent renewal of the fault movements. As evidence of fault origin, many escarpments closely resemble fault scarps on the land. The finding of trenches along the base of slopes, particularly in the Pacific, is strongly indicative of faulting, since a slope built up of sediments could not have these depressions last because they would soon be filled by sedimentation. The large number of earthquakes that occur on or near the continental slopes adds more weight to the argument. Even in the Atlantic one of the greatest earthquakes of all time occurred on the continental slope south of Newfoundland, causing the cable breaks that were referred to previously. Finally, the continental slopes in some places are cut diagonally across the trends of land structures. This is clearly indicated in California south of Carmel, where the land mountain ranges come in diagonally to the coast and are cut off by a submarine escarpment. Despite these indications of faulting, it is now evident that many of the slopes have been built forward by sedimentation.

Fortunately for the residents of coastal areas, most of the fault scarps of the sea floor are not now active. If, for example, the great fault scarp[6] off west Florida should have renewed faulting along it, tidal waves might develop that would be disastrous to the low coastal areas where large beach colonies are located, such as St. Petersburg. In fact we on the coast of California can be thankful that the submarine faults, which are so numerous, either are no longer active or are accompanied by horizontal, rather than vertical, movements. No appreciable tsunamis are known to have occurred on this coast, although small waves from the distant Aleutian faults raise and lower the sea level slightly, causing very minor damage.

Slope Preservation by Landslides

The renewal of movements on submarine fault scarps along the continental slopes prevents many of these slopes from becoming great sediment embankments, but of equal importance are the submarine landslides and the turbidity currents, which apparently are set up by the slides. In areas where faulting is of minor importance, the slides appear to be a major factor in preserving the relative steepness and irregularity of the slopes. Shortly after oceanic cables were laid, they began to have breaks, and most of these breaks were found to occur on the continental slopes. As early as 1897 an English engineer, John Milne, described many of these breaks and attributed them to landslides. As discussed previously, in recent years a number of scientists, including Maurice Ewing and Philip Kuenen, have come to believe that the cable breaks, such as those related to the Grand Banks earthquake (Fig. 9), are due primarily to turbidity currents. We need more definite information about turbidity currents and other processes operating on the sea floor, but at present it seems more likely to me that some kind of landslide or progressive liquefaction of slope sediments after an earthquake is the explanation.

As the delta of the Mississippi River is built forward onto the continental slopes, small gullies develop (Fig. 54). These have characteristics of landslide scars such as basin depressions and mounds. If currents were their cause, the sediments on the valley floor would consist of concentrates of the coarser slope materials, but cores that we took from the valleys and from the intervening

---

[6] It may not be a fault scarp despite its appearance.

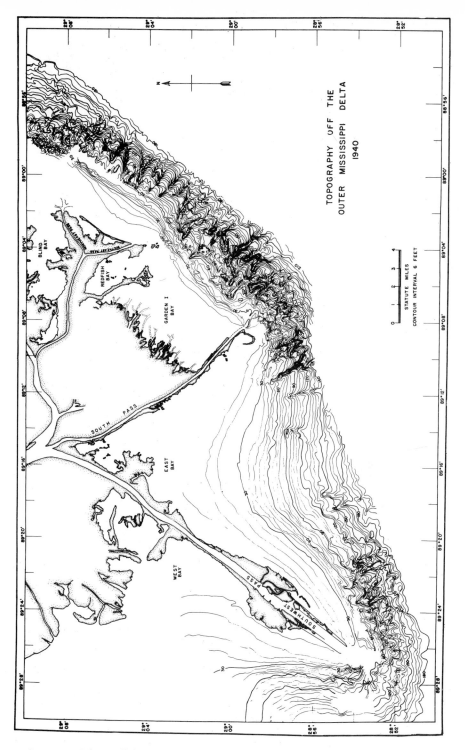

Figure 54. *The small landslide gullies that move forward with the advance of the Mississippi Delta over the continental slope. They do not extend down the slope deeper than 200 feet.*

ridges showed similar sediments with no indication of valley currents. According to Karl Terzaghi, landslides on such low slopes are possible only where there is very fine saturated sediment being deposited rapidly so it cannot become compacted to any extent. Fine sediments deposited more slowly on steeper slopes, however, are also subject to landsliding, and some of these slides may indeed produce turbidity currents that carry the sediments away from the slope and deposit them on the deep-ocean floor or in basins on a lower portion of the slopes.

Where the continental slopes are very steep, as on the escarpment off west Florida, the gradual creep under the force of gravity undoubtedly produces large cracks in the slope just as it does along the coastal cliffs cut back by the waves or on the steep sides of land canyons. Very likely some of the jagged appearance of continental slopes shown on fathograms can be accounted for in this way.

It is very difficult to obtain good reflection profiles showing the structure under the steep slopes. However, there is some clear evidence to confirm the fault origin inferred for many of these slopes. On more gentle slopes, the profiling equipment has revealed many huge landslide blocks, and small faults are seen elsewhere.

CHAPTER VII

CANYONS OF THE SEA FLOOR

The great canyons that are cut into the submarine slopes of the world in so many places provide one of the most controversial subjects in marine geology. Judging from soundings, these canyons resemble river-cut canyons, but why should river valleys extend down all of these submarine slopes beyond the depths where rivers can cut?[1] When submarine canyons were first described almost a hundred years ago, it was assumed that they were old river valleys that had sunk below the surface of the sea, the converse of mountain ranges that have been raised up out of ancient seas. This may sound like a perfectly straightforward explanation, but geologists have been attacking it now for the past seventy-five years. Well-known scientists have suggested almost a score of alternatives. Before we get embroiled in this controversy, we can profitably investigate some of the facts (regrettably still rather scarce) and consider a few of the necessary definitions.

Defining Submarine Canyons

The term *submarine canyon* has unfortunately been used to cover a number of rather distinctly different types of elongate depressions or valleys on the sea floor. These include (1) the broad-floored and steep-walled troughs like those of the Aleutian Island submarine slopes (Fig. 53), (2) some elongate depressions of slight depth below surroundings that extend for miles along the deep-ocean floor, (3) the small gullies that cut the forward-building submerged slopes of great deltas (Fig. 54), and (4) the winding V-shaped valleys with many tributaries that extend down most continental slopes. These last are the features to which the name submarine canyon will be applied in this book. They are, after all, the only valleys of the sea floor that are comparable to the great canyons of the land surfaces.

Description of Some Well-Explored Submarine Canyons

Not many submarine canyons have been studied sufficiently or even well enough sounded so that they can be described except in very general terms. By far the best known are those located in southern California directly off

---

[1] Actually large rivers flowing near the sea are capable of cutting holes several hundred feet deep in their channels. Near the mouth of the Congo there is a hole cut 400 feet below sea level, but this is only a small local basin.

Scripps Institution. Other areas of which there is fair knowledge include canyons off Cape San Lucas, at the tip of Lower California; the canyons off Carmel and Monterey bays, California; some of the canyons off the French Riviera and Corsica; a canyon at the mouth of the Congo; and a few canyons along the coast of Japan. In these places the description is based in part on scuba or deep-diving vehicle, on samples of sediment from the bottom and of rock from the walls, and on numerous sounding lines.

The canyons with their rocky walls terminate at the foot of the steep continental slopes, where there are great fans of sediment. A fifth type of valley, quite different from the canyons, is often found crossing or partly crossing these fans in continuity with the lower end of the canyons. These features, called *fan-valleys,* are cut into the unconsolidated sediments of the fans. Most of these have little depth below the surrounding fans, but may have steep slopes on the outside of their curving courses. Most of them have natural levees of sediment built up along their sides, rising above the adjacent fan, just like the natural levees that are built above the sides of the river channels, which run through the great deltas of the world. Also, like deltas, the fan-valleys may have distributary channels and they lack the tributaries characteristic of submarine canyons.

La Jolla canyons. It is pure chance that the Scripps Institution of Oceanography happens to be located so close to two submarine canyons that from the end of the Scripps pier you can actually shoot an arrow into one of them (Fig. 55). Scripps Canyon, as we have called it, heads directly off the beach a little north of the institution and runs diagonal to the coast for a mile, where it connects with La Jolla Canyon, which starts its precipitous drop in a series of gullies located about 1,000 feet from the La Jolla Beach and Tennis Club, nearly a mile south of Scripps pier.

If you were to don scuba gear and swim down the head of Scripps Canyon, you would start your descent in a shallow sand chute where the water is only about 15 feet deep. As you swam down the axis of the little valley, the walls would get higher and you would see great masses of kelp and eel grass along the floor. At a depth of 50 feet the valley walls steepen until a rock gorge is encountered (Fig. 56). Looking up from the floor of this gorge, you would see that the walls were overhanging in places and almost vertical elsewhere. Schools of fish would swim past you as you proceeded.

At 100 feet you would pass a steep entering tributary, and just beyond you would see several hanging valleys nicking the top of the wall as they enter the main valley. If you could continue to a depth of 175 feet (without getting the bends), you would see the juncture with a main tributary. Following up this tributary you would have to pass through a gorge so narrow that with extended arms you could touch the two walls as you swam through. Higher up, the canyon widens a little, but the walls remain precipitous until this tributary also heads into a sand chute.

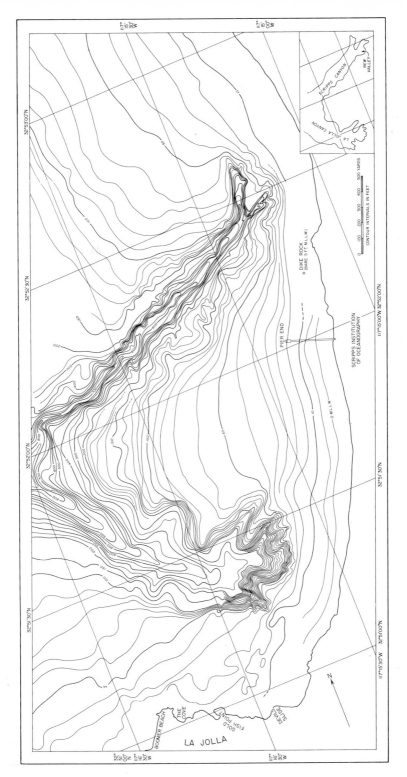

Figure 55. Contour map of Scripps and La Jolla submarine canyons.

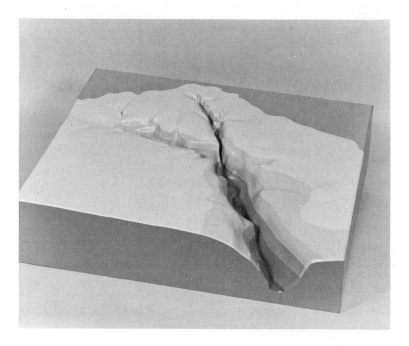

Figure 56. Sumner Branch at the head of Scripps Canyon. Note the short tributary that enters Sumner Branch as a hanging valley. The model was based on contours and was constructed by the Navy Electronics Laboratory. Photo courtesy of R. F. Dill.

Beyond the juncture of the two tributaries the floor of the canyon has now been explored by us at Scripps Institution in Cousteau's *Diving Saucer*. We were amazed to find that the vertical walls, seen by scuba divers at shallower depths, not only continue all the way to the juncture with La Jolla Canyon, but often have overhanging ledges. In one place, the gorge was so narrow that while we followed down along one vertical wall watching it through the port-holes (both on the same side), we were alarmed to feel a jar at our feet which resulted from the *Saucer* hitting the other wall, nine feet away. Tilting the *Saucer* on its side, we could look down and see that the cut was too narrow for further descent. During these dives, we found that the axis of the canyon, where it could be reached, had an irregular steplike descent with vertical drops interspaced with rather gentle slopes. In places, the boulders had clearly fallen from the walls in very recent times, as we could tell by their lack of a covering of organisms that characterizes most of the walls. Also, we found some of the lower portions of the canyon walls were smoothed as if they had been cut by glaciers, having the polish and even the grooves that characterize glaciated surfaces (Fig. 57). This smoothing and grooving has been explained by Robert

Figure 57. Sandy floor and vertical wall of Scripps Canyon at a depth of 460 feet. Note smoothness of wall and partially truncated pholad borings (round holes). The erosion is due to the creeping mass of sand that recently covered this lower portion of the wall and has subsequently been carried away. Photo from Cousteau's Diving Saucer *by D. L. Inman.*

Dill, of the Navy Electronics Laboratory, as the effect of creeping sediment, which is bound together with strands of kelp and marine grass and contains within it many rocks that are pressed against the walls as the sediment creeps downward under the influence of gravity. It had been observed previously in the shallow canyon heads.

Beyond the juncture with La Jolla Canyon the walls are somewhat less precipitous, but the valley continues down the slope until it encounters a gently inclined fan that extends into the bottom of San Diego Trough at about 3,000 feet below sea level.

Below the rock-walled canyon, a fan-valley winds across the fan that is built into San Diego Trough. This has also been explored by a deep-diving vehicle, in this case the bathyscaph *Trieste*. Most of these dives have been made by scientists of the Navy Electronics Laboratory. I was able to join them in one descent. The fan-valley has some very precipitous walls, even as much as 75° being reported. The floor has a thin mud cover in most places but is directly underlain by sand, and even gravel was found locally. At one place, Robert Dill observed a series of cobbles along the floor of the valley, apparently a part of the formation making up the fan. Terraces along the sides of the valley are quite clearly due to landslides, very much like the land terraces found along the base of the cliffs.

Cores taken along the axes of the La Jolla canyons have yielded sand layers and some gravel at a variety of depths (Fig. 58). In some cases, only sand was obtained, but in general the sand was found to alternate with layers of mud along the length of the cores. The canyon sand is as free from mud as the sand on the long beach that runs south from Scripps Institution. One core had a lump of coarse sand that compares with the beach sand at La Jolla Cove, south of La Jolla Canyon. Not only does the sand occur in the canyon, but it is even found on the floor of the adjacent fan and even in the trough out beyond the fan-valley.

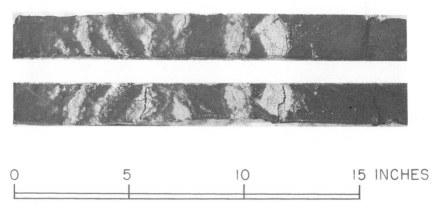

O   5   10   15 INCHES

Figure 58. The two halves of a core with irregular layers of sands obtained from the floor of La Jolla Submarine Canyon at a depth of 2,100 feet. Sand layers (light portion) are shown alternating with deep-water clay layers (dark portion).

Monterey submarine canyons. The canyons that run out of Carmel and Monterey bays in central California completely eclipse those of La Jolla in their gigantic dimensions. Here we have a sea-floor valley that, at least along a portion of its length, is really comparable with the Grand Canyon of the Colorado (Fig. 59). The dendritic tributaries that go to make up this underwater valley system are very much like the canyons that cut the mountain slopes on the adjacent land (Fig. 60).

Carmel Canyon has several branches, one directly off the Carmel River valley, but the branch that penetrates deepest into the land connects with San Jose Creek canyon, running seaward just north of Point Lobos [2] State Park (Fig. 61) This canyon head comes in so close to the beach that you could throw a pebbl into it from the water's edge. The submarine canyon is cut in granite just like the adjacent land canyon, and in fact the two have very much the same shape. If the land were uplifted, no one would be puzzled by the appearance of the

[2] Lobos is Spanish for sea lions, rather than wolves, in a coastal area.

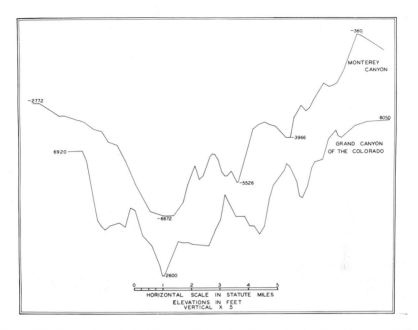

Figure 59. Comparison of the Monterey Submarine Canyon to the Grand Canyon of the Colorado River. Profiles drawn at the same scale and based on the same number of observations.

portion that is now under water because it would look just like an ordinary land canyon.

The head of Monterey Canyon is something different. If that were uplifted, a marked break would be seen between the flat Salinas Valley and the steeply inclined canyon that begins just beyond the present beach. The group of sand chutes that are found at the canyon head connect at a depth of 400 feet a mile from the beach, and from there on a sizable canyon winds out of Monterey Bay and is joined by various tributaries and finally by the deep Carmel Canyon, which enters at 6,600 feet below sea level. Dredging of the walls of Monterey Canyon has yielded sedimentary rock of Pliocene age, probably underlain by Miocene rock and then on one side by granite. Cores taken along the valley floor here have again yielded a surprising amount of sand in view of the great depths involved. One core taken at 4,800 feet consisted of pebbles at the bottom, sand higher up, and finally mud at the surface, a relationship usually referred to as *graded bedding*. At the head of the canyon, Frank Haymaker, formerly a diver for the United States Navy, descended into the sand-chute valleys and obtained some cores of the bottom for us by driving a pipe in with a hammer. The cores consisted of alternation of coarse sands, like those of the nearby beach, and muds, the same as those found off La Jolla.

121

Figure 60. Showing the axes of the canyons that extend out of Monterey and Carmel bays and southward from Point Sur along the California coast. It will be seen that these compare closely in outline with the land canyons in the vicinity.

The outer terminus of Monterey Canyon appears to lose its V-shape at a depth of about 9,500 feet, where it runs into a sea-floor trough, which in turn continues to about 10,500 feet. Beyond the trough, there is a relatively narrow fan-valley with natural levees. At a depth of 10,400 feet, this valley has a large

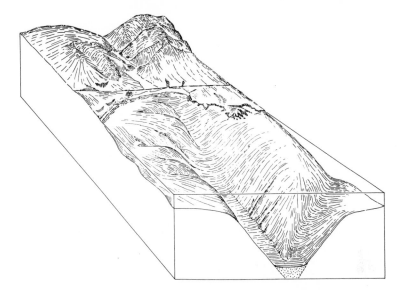

Figure 61. Block diagram by D. B. Sayner showing the general relationship between San Jose Creek Canyon, California, and the submerged head of the Carmel Submarine Canyon. The thin black line represents the surface of the sea.

meander, almost returning on itself after a course of 20 miles. It must resemble one of the giant entrenched meanders in the San Juan Gorge of the Colorado Plateau. The fan-valley then continues for another 80 miles before it terminates in a large fault valley at a depth of 11,500 feet.

Canyons off Lower California. Because of their proximity to the Scripps Institution of Oceanography, Lower California and the Gulf of California have been the focus of a number of Scripps Institution cruises. On four of these we surveyed the canyons off the lower tip of the peninsula. One canyon extends into the entrance of the harbor at Cape San Lucas, where it is flanked by a granite promontory. Since 1962, Robert Dill and other scuba divers have been diving frequently into the shallow canyon head. Finally, in 1965, we brought Cousteau's *Diving Saucer* to the area, and I had the opportunity, as did Dill and D. L. Inman, to follow the valley down to a depth of a thousand feet. These dives have shown us that the canyon is cut into granite and other hard rocks. The canyon has many small tributaries coming in mostly as hanging valleys. On the south side, the valley runs along a rocky point. The adjacent canyon wall is very precipitous. Sand becomes stored on the steep upper slope and from time to time moves as sand flows (first photographed by Conrad Limbaugh in 1959) that turn into sand falls where the mass goes over vertical cliffs. The floor of the canyon below these falls has been peppered with angular rocks that must be broken off, at least in part, by the sliding and falling sand.

The canyon floor also has extensive ripple marks, showing that currents of some force are moving up and down the canyon.

Traced outward by echo soundings, San Lucas Canyon (Fig. 62) was found to wind down into deep water and to 7,200 feet, where it ends in a large submarine fan. Like other typical submarine canyons it has a number of tributaries entering it from both sides.

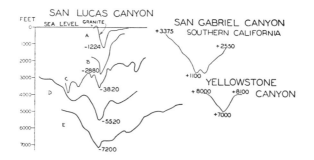

Figure 62. Sections of submarine canyons off the southern tip of Lower California compared with land canyons. All sections have the same scales.

Perhaps the most interesting thing about the Lower California submarine canyons is that almost all of them are clustered around the lower tip of the peninsula. Sounding lines to the north along the steep submarine escarpments bordering the peninsula on the gulf side showed no valleys of any consequence. On the Pacific side there is a little valley half way up the peninsula, just south of Cedros Island. Another bowl-shaped valley off Ensenada in Todos Santos Bay hardly warrants the use of the name canyon. The concentration of canyons at the end of the peninsula might be related to the rainfall which is greater in that area than along the rest of the peninsula, where there is a true desert climate.

East coast canyons. In their general appearance the east coast submarine canyons (Figs. 63, 32) do not differ particularly from the California and Lower California canyons. The great distance from shore and the considerable depth of water at the canyon heads make it difficult to obtain details. With the present type of echo-sounding device, for example, it is impossible to find vertical walls or gorges like those off Scripps Institution (Fig. 64). Henry Stetson, however, was led to suspect the existence of vertical cliffs in a canyon off New England because of the appearance of a huge block of sandstone that he dredged. This rock had been deeply gouged at right angles to the stratification, indicating that the wire was pulling the dredge up vertically along the front of a projecting sandstone cliff. The rocks that Stetson found in his exploration contained fossils, many of them as old as the Cretaceous (about 100,000,000

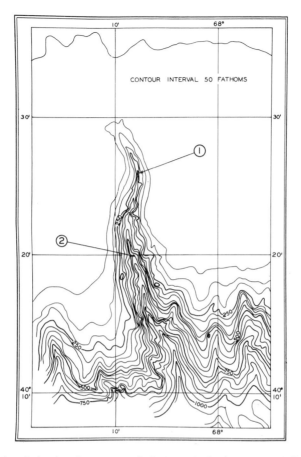

Figure 63. A typical submarine canyon off the New England coast. (1) indicates a point where Cretaceous rock was dredged, and (2) indicates the position of a core in which glacial marine sediments were found under postglacial sediments.

years back in the time scale). The sediments found on the floor of the east-coast canyons indicate a long period of slow accumulation. At the bottom of a six-foot core Stetson found foraminifera of a type that are now living in the upper waters of the sub-Arctic, and therefore it can be assumed that this lower part of the core represents a glacial stage (10,000 years or more in the past) when the cold waters existed this far to the south.

The canyon off the Hudson River (Fig. 32) is remarkable for its large number of tributaries, although each of them is quite short. The walls have a maximum height of about 4,000 feet. Dredgings by Stetson failed to yield rock, Maurice Ewing and his Lamont Observatory group found Miocene clays in the fan-valley. One large block of rock was obtained from the floor of the canyon at a depth of 7,000 feet. Coring at a depth of 12,000 feet, they found that the

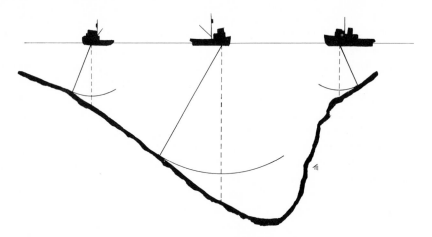

Figure 64. The sound sent out by an echo-sounding machine echoes from the nearest reflecting surface, thus giving a false impression of the true depth directly beneath the vessel.

sediment in the canyon consisted of gravel with shells and clay cobbles. Ewing's group traced the canyon outward, finding that it continues as a typical fan-valley running across a great fan that has been built into the western basin of the Atlantic. Including the inner canyon and the outer fan-valley the entire length of this marine valley is about 180 miles, and the terminal depth is nearly 14,000 feet. Samples from the outer fan-valley and from the surrounding fan have included great quantities of sand, much of it washed clean, like the sands around the base of the La Jolla Canyon.

Canyons along the coasts of the French Riviera and Corsica. The canyons of the Mediterranean are in many respects the most intriguing in the world. Here the slopes have more canyons than in any other areas that has been surveyed. These valleys extend well below the sill depth of the Mediterranean, which is at 1,400 feet. Many of the canyons head right into the coast. The water is so clear in some places that you can actually look down in the canyons from land. I stood on the platform at the top of an old Genoese tower and looked down into the heads of two tributaries of the huge canyon that runs out of the Gulf of Porto on the west coast of Corsica. Walking along the coast near the Eden Hotel at Eze, west of Monaco, I could see the steep wall of a submarine canyon disappearing at a depth of about a hundred feet below the waters of the Mediterranean. My friend Robert Dill, a marine geologist of the Navy Electronics Laboratory, dove with André Portelatine for 200 feet into one of the canyons near Monaco. He was impressed by the narrow rocky gorges that he encountered and compared them with the river-cut gorges of the Alpes Maritimes above water on this same coast. On the other hand, descents in a bathyscaph by Cousteau and Georges Houot of the French Navy into much deeper canyons off Toulon disclosed only mud-covered walls.[3]

In one respect the canyons of the French Riviera contrast with those of West Corsica. As an example of the former we can refer to the situation off Nice. While accompanying Professor Jacques Bourcart of the Sorbonne on the French Navy vessel *Elie Monier,* I watched the fathometer tracing a series of sharp gorges and narrow ridges as we ran along the straight shore of the famous resort so close that we could see the Bikini bathing suits on the beach. Here, as in Monterey Bay, the topography on land is flat, but the slope is carved into valleys just beyond the coast. These valleys become large canyons farther out. Here dredging has brought up rock formations described by the French as representative of various ancient geological periods. In our work off Nice we obtained canyon-floor cores that had sand layers like those from the American submarine canyons. Dragging a sled with an attached camera (designed by Jacques Cousteau), M. Gennesseaux of the University of Paris took extensive photographs of the bottoms of the canyons off the French Riviera around Nice. These photographs showed that ripple marks were common in the canyons, and masses of cobbles were scattered along the canyon floors.

The Corsican canyons, unlike those off Nice, fit right into the pattern of the land surface (Fig. 65). Each bay on the west coast has its submarine canyon, and in fact each tributary bay has a continuation beneath the sea. This relationship is even more pronounced than that found at Carmel in California. It is just as if a mountain range had been submerged recently, and part of its canyons were under water and part above. In fact one can scarcely doubt that this is actually the case. Napoleon would have been surprised to learn that the Bay of Ajaccio, where he spent his youth, had its various land canyons continuing on beneath the sea.

The Congo Canyon. The Congo has the greatest flow of any river in Africa, yet it has no delta and in fact enters a deep estuary at the port of Banana. This estuary is the head of the 145-mile-long Congo Submarine Canyon (Fig. 66). For those who might suggest that a recent great submergence has drowned the delta and the lower reaches of the Congo River valley, the fact has to be considered that the wide continental shelf found both north and south of the Congo estuary terminates at about 420 feet, which is close to the average for the shelves of the world. Furthermore, except at the Congo estuary, the coast is straight rather than deeply indented like the west coast of Corsica. At the entrance of the estuary, the water is almost 3,000 feet deep. If this were a glaciated country, the deep bay would not be difficult to explain, since similar depths occur in glacially excavated fiords, but instead it is found in a hot climate near the equator. Furthermore, the bottom of the estuary, unlike a fiord,

[3] Georges Houot and Pierre Henri Willm, *2,000 Fathoms Down* (New York: E. P. Dutton & Co., 1955).

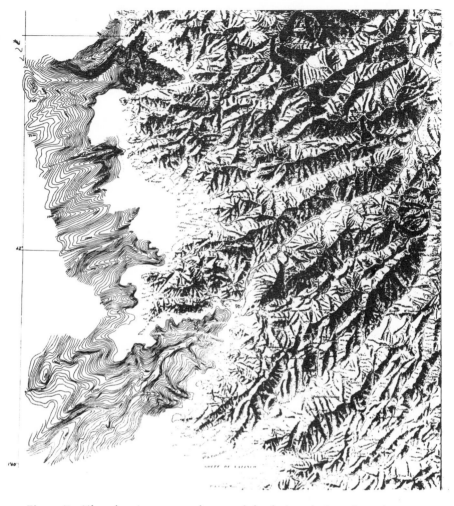

Figure 65. The submarine canyons that extend deeply into the bays along the west coast of Corsica. The clear relationship between the submarine canyons and the land canyons can be observed. Contours by Jacques Bourcart of the Sorbonne.

slopes outward almost continuously, as far as one can judge from the fairly abundant soundings.

The outer part of the Congo Canyon has recently been surveyed by Heezen of the Lamont Geological Observatory who discovered an outbending fan with shallow fan-valleys extending beyond the canyon, as is the case off other great submarine valleys. Many distributaries occur in this fan-valley.

Canyon of Tokyo Bay. A submarine canyon penetrates for 12 miles into the mouth of Tokyo Bay. This canyon has a much more winding course than the

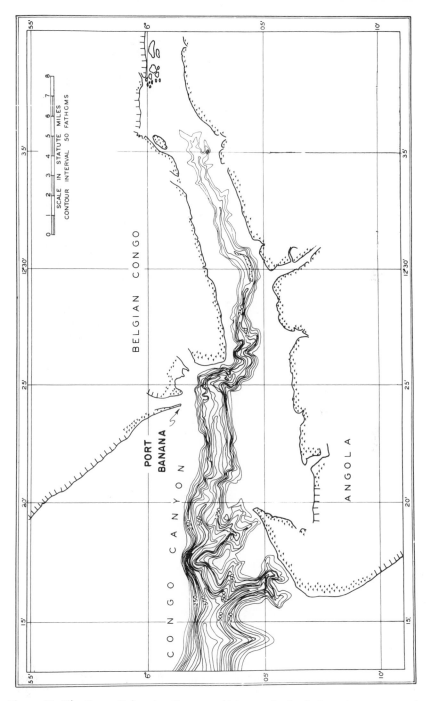

Figure 66. The Congo Submarine Canyon, which extends deeply into the estuary at the mouth of the Congo River. Data from Portuguese and British charts.

Congo Canyon. The tributaries are also better developed; in fact it is like the canyons in the Corsican bays except that it heads in a broad shallow bay. Much sampling by the Japanese, and more recently by our Scripps expedition, has shown that there are rocky walls. Canyon-floor samples include much sand.

The Tokyo Submarine Canyon extends out to a depth of about 4,620 feet where it encounters an elongate straight-walled trough coming out of Sagami Bay. This trough has clearly been caused by faulting, and it was in fact along this trough that the movements occurred that produced the disastrous Tokyo earthquake of 1923. The association of the winding canyon and the straight-sided fault trough is one of the puzzles, the answers to which may contain some of the clues of submarine-canyon origin.

A small canyon off Bōsō peninsula, south of Tokyo Canyon, was explored in 1961 by our Scripps expedition. We obtained photographs along the floor showing alternately rocky bottom and large steep-sided ripple marks, both indicative of strong currents.

Canyons or troughs associated with great deltas. The deep valleys that penetrate the continental shelves off several large deltas may not belong in this discussion of submarine canyons because they have few of the river-canyon characteristics possessed by the submarine canyons that have been described. The Indian shelf is the place where the delta-front troughs are best developed. The Swatch of No Ground off the Ganges Delta is a trough-shaped flat-floored depression that penetrates across the shelf to within 20 miles of the delta margin. At the edge of the shelf it has a depth of about 3,000 feet. It continues seaward down the Bay of Bengal slope as a typical fan-valley with well-developed natural levees. In 1963, the U. S. Coast and Geodetic Survey's Indian Ocean Expedition traced the valley south, where it was found to branch into three or more distributaries and to show a very complicated pattern that was impossible to chart accurately because of cloudy weather and no electronic method of navigation (satellite navigation was not yet established). Widely separated sounding lines suggest that this valley system may continue as far south as Ceylon.

The Indus Swatch heads within three miles of the shoals that surround the Indus Delta (Fig. 67). From there it extends seventy miles to the edge of the shelf as a relatively straight trough with a floor that slopes outward continuously. At the shelf margin it has a depth of 3,720 feet. All transverse profiles show a fairly broad flat floor. Off the western part of the delta of the Mississippi River there is a similar trough-like valley, but this one heads fifty miles from the shore. Geophysical measurements on the shelf, however, have shown that the trough can be traced in to where it connects with a buried valley that has a floor 400 feet below sea level at the delta margin. The submarine trough can be traced seaward to a depth of more than 6,000 feet, where it appears to terminate and has a great fan as a seaward prolongation.

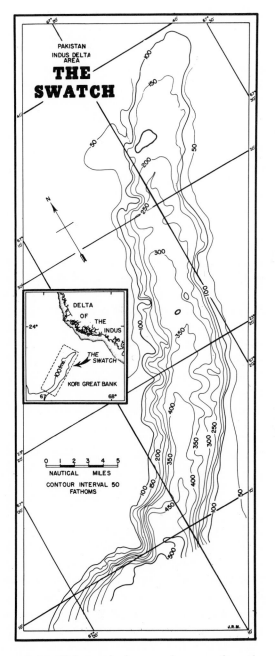

Figure 67. The straight-walled trough that extends across the wide continental shelf adjacent to the Indus Delta. From a recent survey by the Pakistan Navy.

As far as we know now, none of the trough valleys have been cut through rock, and it seems likely that they merely penetrate deltaic sediments. None of them have clear evidence of tributaries, nor do they have the sinuous course of typical submarine canyons. The sediments of the Mississippi Trough are known to contain sand layers, and shallow-water faunas occur at depths of as much as 4,800 feet.

Origin and Preservation of the Canyons

In attempting to explain the submarine canyons it is important to use type examples that fit the definition given early in the chapter; that is, valleys that resemble land canyons in extending down the regional slope and in having winding courses, a general V-shaped cross section, and many entering tributaries. This covers all the canyons described above except the fan-valleys and the troughs off deltas. Including other types needlessly complicates the explanation. Furthermore, even land valleys are not all explained in the same way. Some of them are primarily cut by rivers and slope wash; others owe their origin to glaciation, to faulting, or to landsliding. Thus the question at issue is what is the primary process that has led to the existence of submarine canyons.

Complicating factors. In view of the preceding descriptions of individual canyons, you may have wondered why so many geologists have not been happy with the idea that the canyons are drowned river-cut valleys. If the sea-floor valleys all had the relation to the adjacent land valleys that is found on the west side of Corsica, there would probably be an overwhelming adoption of the drowned-canyon hypothesis. Corsica, however, is an exceptional case. Virtually all of the other submarine canyons form a topographic discontinuity with the valleys of the adjacent lands. The majority of canyons are located off fairly straight coasts that do not look at all as if the area had been sinking. They do not have the embayed valleys that are certain to form as the result of the submergence of an area with hills and valleys. Also the submarine canyons almost always have far steeper gradients than the adjacent land valleys.

Another serious difficulty is that, as far as we know, the submarine canyons do not stop part way down the continental slopes but continue as valleys to the bottom. If cut by rivers, why does not the valley stop at the lowest level at which the river was flowing when the slope became submerged (Fig. 68A)? Perhaps an even more common objection offered to the river-erosion idea is that the canyons are world-wide. When we have better surveys, this last objection may disappear because most of the examples of submarine canyons that have been shown on maps are not clearly established as fitting the rather restricted definition that is given here, and therefore many of them may have been formed by an entirely different process from that which formed what we are here calling true submarine canyons.

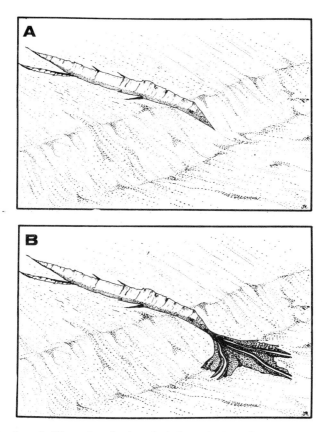

Figure 68. Sec. A: Illustrating the hypothetical termination of a submarine canyon part way down the continental slope. Sec. B: Showing the actual relationship of the submarine canyons to the continental slope with valleys continuing beyond the excavated portion of the canyon down the floor of the deep fan.

Old discredited hypotheses. In attempting to avoid the difficulties confronting the river-excavation hypothesis, scientists have offered surprisingly numerous explanations for the origin of the canyons, all of them related to processess that can operate on submerged slopes. Most of these ideas have now been abandoned because they appeared to raise more problems than they explained, but it will be just as well to review them briefly and to remember that, in science, discredited hypotheses frequently have an embarrassing way of coming back to life.

Several geologists have thought of faulting on the sea floor as the cause of the canyons. The dropping of blocks of the earth's crust explains many land valleys, and this process almost certainly accounts for some of the sea-floor valleys, such as those in the Aleutian area described previously (Fig. 53). The nature of the typical submarine canyons, however, is not at all comparable to that of fault valleys. The winding courses, dendritic tributaries, and V-shaped cross sections apparently rule out faulting as a major cause.

Several authors have thought of underground circulation of water with the solution of caves and the collapse of the cave roofs as the principal cause. Ground water, however, will not circulate actively below sea level unless it has a great head of water to drive it forward against the pressure of the sea water. This requires high land and layers of rock dipping seaward under the ocean. This is not usual inside canyons. Furthermore, the collapse of caves should leave very irregular blocked valleys, and we have found these only along the coast of the Gulf of Mexico, where there is not enough relief on land to cause active ground-water circulation.[1]

Various types of currents have been suggested, mostly by geologists who were not very familiar with oceanographic investigations. Such currents as undertow are known to go contrary to the dynamics of sea water, and upwelling, which brings water up from the depths, is known to be a very feeble flow, quite powerless to cut great gorges or any other type of valley. Probably the same is true of downward-moving currents caused by chilling of sea water along the shore or evaporation in partially closed bays, both of which cause the water to become more dense than that found on the adjacent slopes. Tsunamis or other great sea waves have been suggested, but it is doubtful if they produce very appreciable currents except in shallow water near the shore. Nor does the distribution of canyons fit the areas where tsunamis are common. Landsliding is an interesting hypothesis, but slides are usually considered as only a partial cause of the canyons. Landslides are certainly capable of producing valleys in soft unconsolidated material more easily than in hard rocks such as granite. More will be said later about landslides as a co-operative cause and as a means of preserving the canyons.

Turbidity currents. The only hypothesis other than river erosion that is widely held by geologists at the present time is that the canyons were cut by turbidity currents (see pp. 23-26). Originally this explanation was suggested by Reginald Daly of Harvard University, but it was not given wide support until Philip Kuenen demonstrated in his 100-foot tank at Gröningen University in Holland that sediments of sand size and even gravel could be carried in suspension along the sloping floor of the tank when a homogenization of water and sediment, including considerable mud, is poured into the shallow end of a tank. More support came from the work of Maurice Ewing and his Lamont Observatory group when they reported the widespread occurrence of sand deposits in the fans around the base of some of the submarine canyons. Then came the report by Bruce Heezen and Ewing on cable breaks accompanying the Grand Banks earthquake, which they interpreted as due to 60-mile-

---

[1] This contention is challenged by recent discoveries of large underwater springs off the low Florida coast.

an-hour currents. If sea-floor currents can flow at that speed and can break strong cables, they should certainly be capable of excavating canyons even in granite. However, as indicated previously, the evidence is more favorable to a much slower speed.

Unfortunately, we have not reached the end of the trail in this controversy. Karl Terzaghi's theory of progressive liquefaction of slope sediments following an earthquake (see p. 25) seems to present an alternative to that of the supposed high-speed turbidity currents. We have yet to prove that there are high-speed turbidity currents. The only speeds measured so far are in artificial lakes where the forward motion of the turbidity currents is not known to exceed one mile an hour. The deposits left by turbidity currents are all indicative of slow speeds even in the Grand Banks cable-break area. We are planning to set up equipment for measuring these currents in the submarine canyons off Scripps Institution, but as yet our devices have not been successful.

Some indirect lines of evidence appear to throw light on turbidity currents as they operate in the canyons off southern California. The well-sorted sand layers found in cores taken from the bottom of the canyons show that appreciable current has passed along the canyon floor in order to transport and sort the sand. On the other hand, the finding of alternating sand and deep-water mud layers [4] does not indicate that the currents introducing the sand were capable of causing much erosion, since otherwise the soft muds under the sands would have been cut away and, as the current slackened, the sand would have been deposited directly on the rock formations into which the canyons have been cut. It seems more likely that the currents are capable only of transporting sand and other sediments and not of eroding the bottom. The same conclusion appears to come from the investigation being made by Kuenen and others of ancient turbidity-current deposits. In one place, the coarse sediments of the turbidity current were found on top of fine deep-water sediments in which there were worm tracks that had not been obliterated by the current. On the other hand, results of current action such as ripple marks and cross-bedding have been discovered in other supposed turbidity-current deposits of the ancient sediments. The evidence is not yet complete.

River erosion at remote periods. At one time I thought that the best explanation of the canyons was that they were cut as a result of a great lowering of sea level during some early stage of glaciation when the ice caps might have been much larger than in the last, or Wisconsin, stage. Some years ago I found that this idea had to be abandoned. The exploration of some of the flat-topped seamounts had shown that there could not have been any great lowering during even the earlier stages of the glacial period. The shallow-water fossils found on these seamounts are much too old. Furthermore, the fact that the submarine canyons are so generally discordant with most of the coastal valleys inside them

---

[4] Known to be deep-water deposits because of the type of foraminifera.

helps rule out any great submergence in Pleistocene or more recent times. Finally, buried canyons at the canyon head contain rocks as old as Miocene (about 20 million years ago).

There should not be such serious objections to the idea of submergence of canyons at various times in the remote geological past. The findings of geologists working on the sedimentary rocks obtained from deep drilling along the continental margins show that over the course of long periods of time great submergences have taken place. Almost every coastal valley that is inside a submarine canyon is underlain by thousands of feet of sediments containing faunas indicative of deposition at a much higher level than now exists at the place from which the rock was derived. For example, borings on the east coast show that Cretaceous rocks deposited on land now lie at a depth of 6,500 feet below sea level. In the Bahamas, well drillings have obtained rocks of shallow-water origin at a depth of 14,000 feet. At the head of the Monterey Canyon, wells in the Salinas Valley show an old canyon with a base at 5,000 feet below sea level and filled with shallow-water Miocene sediments. In other words, in many places the land margins have stood higher. Perhaps during the time when they stood high many canyons were cut before the submergence allowed the deposition of the thick sedimentary columns.

If the canyons are as antique as indicated by this line of reasoning, there should not be any serious objection to the present land valleys being out of adjustment with the much older canyons of the sea floor. During the millions of years since submergence the lands would have undergone vast changes both in erosion and deposition. This leaves us with the problem of why ancient drowned canyons should be preserved.

Landslides and water deepening in relation to the canyons. The movement of sand along the coast at La Jolla is sufficiently rapid that in a few years the heads of the canyons that penetrate in close to the coast in this area would be completely filled. Yet we have been watching the depth changes in these canyons now for more than thirty years and there has been no net fill. This is explained by the periodic landslides that take place in the canyon heads (Fig. 69). Sand fills in very rapidly after slides have occurred, but after a time (about a year in one head) slides occur and the process starts all over again. The slides take place during earthquakes but more commonly after and during periods of large waves. In 1949, while we were operating our canyon surveys with the help of a United States Army Beach Erosion Board contract, we had a visit from members of the board and were eating lunch in the patio of La Valencia Hotel. Suddenly an earthquake set the table rocking, and we were afraid that the balustrades from the balcony above us might fall on our heads, so we made a quick move away from them. The next time we were able to get out to survey the canyon heads we found that the water had deepened as much as sixteen feet along one of our well-established lines. The material had moved on out,

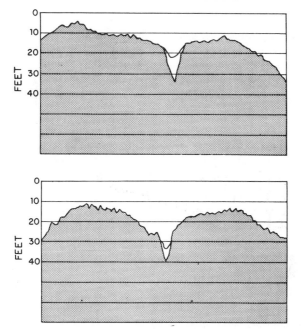

Figure 69. Comparisons of echo-sounding lines obtained in two surveys made in the head of Scripps Submarine Canyon. Note the considerable deepening that has taken place during the interval. These changes are attributed to submarine landslides, and in this case the slide may have been set off by an earthquake.

just as the sand dunes along the coast of Zeeland in Holland slide out occasionally into the North Sea.

At Redondo, near Los Angeles, and at Moss Landing on Monterey Bay, piers have been built out into the edge of submarine canyons. Both of these piers have been damaged by landslides. At Redondo one slide occurred while the pier was being used by fishermen. They found they had to keep letting out more line because of the deepening of the water until it had deepened twenty-five feet and the pier end became so unstable that they had to leave hurriedly. The importance of these landslides in keeping any kind of valley on the sea floor from being filled with sediment cannot be doubted.

Recent Conclusions

Since the first edition of this book was published, extensive investigations have been made of the sea-floor canyons. This work has included scuba-diving into canyons, taking many photographs of the canyon floors and walls, sampling sediments of the floors by using a large box-type of corer, and, most recently, observing the canyons from deep-diving vehicles, such as the Piccard *Bathy-*

scaph, Cousteau's *Diving Saucer,* and the Westinghouse *Deep Star.* Most of this work has been done cooperatively by Scripps Institution and the Navy Electronics Laboratory, but important additional information comes from the French marine geologists.

Among the discoveries coming from this recent work have been: (1) currents other than turbidity currents have been found in canyons that are sufficiently powerful to produce ripple marks, even at the greatest depths; (2) smooth, grooved surfaces have been seen in vertical or even overhanging cliffs along the sides of canyons, which must have been recently eroded; (3) gravel, cobbles, and boulders have been found extensively along the canyon floors, along with evidence that even the boulders are being moved down the canyons; (4) sand layers have been obtained from the canyons with heavy mineral laminae and other characteristics, which are more indicative of ordinary currents than of turbidity currents; and (5) extensive changes in the canyon fills have been detected at depths much greater than the 200-foot limit where such changes had been measured by soundings or observed by scuba divers. These and other discoveries have led to the belief that submarine canyons cannot be explained as merely the result of sinking of river-cut canyons any more than they can be explained as being cut exclusively by turbidity currents. It seems likely that various types of submarine erosion, including the force of gravity producing creep and landsliding, currents due to tides, internal waves, and turbidity currents, have been acting on the valleys for a long time, probably millions of years. This erosion had remodeled and greatly extended various types of slope depressions that may represent, in part, submerged river valleys and, in part, valleys due to earth movements (like Death Valley).

CHAPTER VIII

THE DEEP-OCEAN FLOOR

History of Soundings

The early navigators like Columbus and Magellan had no idea of the depth of the oceans except near the lands. Magellan at least tried to reach the bottom of the Pacific, but he had only 1,200 feet of rope, which of course was inadequate. In 1840, Sir James Clark Ross made the first true oceanic sounding, reaching a depth of nearly 12,000 feet with a weight on the end of a hemp line. This was a tremendous undertaking, and the result did not give the true depth because the hemp must have stretched considerably and because the surface currents must have prevented him from keeping his vessel directly over the sounding weight. In 1870, with the use of piano wire by Lord Kelvin, the stretching of the sounding line was largely eliminated, and soundings became much more accurate. Even today, however, with high-powered electric winches it takes hours to lower weights or instruments to the bottom of the deepest holes of the oceans. With the primitive winches previously available it is no wonder that only some 15,000 soundings had been taken in the deep oceans up to 1923, when echo-sounding machines started to revolutionize these investigations.

Contrary to what might be supposed, instrumental measurements in 1856 actually provided the first indications of the true depths of the oceans. The physicist A. D. Bache computed that the Pacific Ocean had an average depth of about 12,000 feet by using the principle that long-period waves vary in their speed of motion directly with the depth of water that they traverse. He measured the time of submarine earthquakes off Japan as recorded on an early version of the seismograph and used tide gauges on the west coast of the United States to obtain the time of arrival of tsunamis set off by the movements at the time of origin of the earthquake.[1] The interval gave him the average depth, using the formula $C = \sqrt{gh}$, where C is the velocity, g the acceleration of gravity, and h the depth of water.

Among the results of the *Challenger* Expedition of 1872-76 was the confirmation of Bache's estimates of the approximate depths of the deep oceans. The deep-sea soundings by the *Challenger* and other contemporary expeditions unfortunately did not dispel the impression that the deep sea was essentially a flat monotonous plain, although irregularities were found. It was only when the German ship *Meteor* in 1924 began running echo-sounding lines across

---

[1] The earthquake does not produce the sea waves, but both earthquakes and sea waves are the result of faulting of the bottom.

the South Atlantic that it became evident that the deep ocean had great irregularities. Using the *Meteor* soundings, I found that when I compared a profile across the United States with one of the *Meteor* profiles drawn at the same scale they had much in common (Fig. 70). Profiles by the United States Navy across the Pacific showed similar irregularities.

For a long time the United States Navy made routine echo-sounding lines on most transocean voyages. This, however, was not very productive because the echo-sounding devices were not well calibrated and often not well read, and the positions were inaccurate except near shore. As a result, attempts to draw the bathymetry from a plotting of large numbers of sounding lines produced only a nightmare for the compilers in the Navy Hydrographic Office.

One portion of the deep ocean has been rather completely sounded by the United States Coast and Geodetic Survey using well-calibrated machines. This is the area between the United States and Alaska. The ships going back and forth each season always chose new lines, and the result has been a very fine coverage of an important area. Beginning in 1950 the Scripps Institution started an ambitious program of mapping the Pacific. Up to 1957, the Scripps vessels had taken approximately 300,000 miles of sounding lines with well-calibrated instruments, and the coverage, as will be seen in Figure 71, is now getting to be very extensive particularly after the International Geophysical Year cruises. The results of the earlier cruises combined with soundings from other sources have been compiled by H. W. Menard of Scripps Intsitution of Oceanography and have produced much of the information that will be discussed here. In the Atlantic the cruises of the Woods Hole Oceanographic Institution's vessels and Lamont Observatory's *Vema* have also yielded numerous soundings, although the coverage is not as wide as in the Pacific. The International Indian Ocean expeditions, beginning in 1960, have added immensely to the sounding coverage of that ocean.

Topographic Features of the Deep-Sea Floor Defined

The report of the British National Committee on the Nomenclature of Ocean Bottom Features, published in 1953, will be used here for most of the nomenclature of the topographic features of the deep-sea floor. In some cases, however, usage seems to have overruled the committee names, and recent discoveries have added new features.

Rise and *ridge* are defined as elongate elevations on the sea floor having gentle smooth slopes for the rise and steep irregular slopes for the ridge.

Seamounts are isolated sea-floor elevations rising 3,000 feet or more above their surroundings. If these mountains have flat tops, they are called *guyots,*

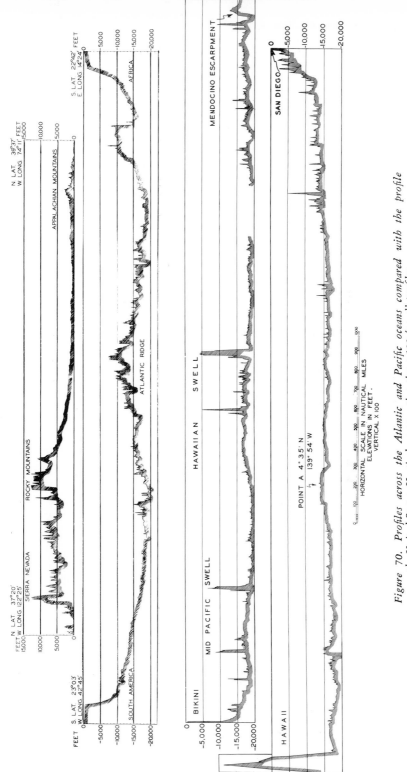

Figure 70. *Profiles across the Atlantic and Pacific oceans compared with the profile across the United States. Vertical exaggeration times 100 for all profiles.*

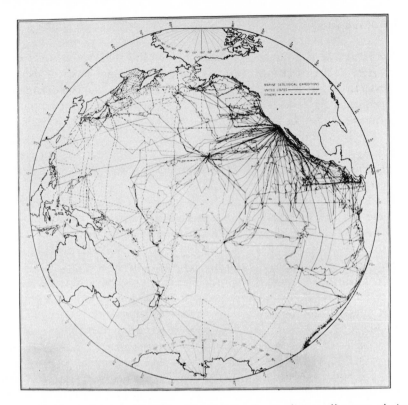

Figure 71. Track chart of expeditions from which echo-sounding profiles were obtained in the Pacific up to 1964. Most of the U.S. lines were run by Scripps Institution. Courtesy of H. W. Menard.

or *tablemounts*. The committee preferred the latter, but usage has ruled in favor of guyots, named after a French geographer.

Sea scarps are elongated and comparatively steep slopes of the sea floor.

Basins are depressions on the sea floor that are more or less equidimensional. *Trenches* are depressions that are long and narrow and at least relatively steep sided. A *deep* is the deepest area of any depression but is not used unless depths are in excess of 18,000 feet.

Deep-sea fans are gently sloping, sediment-covered plains that border the continental slopes in many places. The term *continental rise* has been used for approximately the same thing. *Archipelagic apron* is a term that refers to smooth fans around oceanic islands dominantly constructed of lava flows. They are partly sediment covered.

Deep-sea channels are the elongate valleys that cut slightly below the surface of many of the deep-sea fans and in some cases extend out into the basins.

Abyssal plains are the very flat surfaces found in many of the basins.

Floor of the Atlantic

Mid-Atlantic Ridge. The Mid-Atlantic Ridge (Fig. 72) extends right down the middle of the Atlantic Ocean from Iceland to a point south of the Cape of Good Hope. As H. W. Menard has pointed out, the line equidistant from the continental slopes on the two sides of the Atlantic falls right on top of the Mid-Atlantic Ridge for almost its entire length. The ridge rises about a mile above the deep basins on the sides. A few islands, such as the Azores, St. Paul Rocks, Ascension, and Tristan da Cunha, rise above the surface, but the rest of the ridge is submerged. St. Helena, where Napoleon spent his last days, lies just east of the ridge of the South Atlantic.

The legend of the Lost Atlantis, made famous by Plato, has been endorsed by some romantic geologists and archeologists on the basis of obscure evidence from the Mid-Atlantic Ridge. A little serious study of the source of the legend will show that it is not even supported by the classics. The Atlanteans, according to the Greek writings, lived somewhere beyond the Pillars of Hercules, meaning the Straits of Gibraltar. Any of the islands or even the African coast would do for the legend. The supposed disappearance may mean that the early navigators failed to find the area in their attempt to return to it. Furthermore, the ridge has so far yielded no acceptable evidence of having been above water.

The explorations of the Woods Hole and Lamont groups and of the British National Institute of Oceanography have shown the amazing complexity of the ridge, particularly south of the Azores. Traverses show the presence of as many mountains and valleys as one crosses in flying over the Rockies. On the sides of the ridge there are terrace-like features with flat floors. At first it was thought that these might represent wave-cut terraces cut into a submerging mountain chain, but it was soon found that the terraces are underlain by thick masses of sediment, so that they are now interpreted as deposits formed by turbidity currents in basins blocked by ridges, just as sediments form in the artificial lakes behind dams until the basins are filled up to the level of the dam spillway.

The British appear to have been the first to point out what is perhaps the most amazing feature of the Mid-Atlantic Ridge. Working north of the Azores, they found a steep-sided trench extending along the ridge. It has a flat floor a few miles wide with depths that are almost the same for several hundred miles. This is very similar to the great valley called the Rocky Mountain Trench, which runs between mountain ranges from northern Montana up through British Columbia and finally through the Yukon territory to Alaska. This flat intermontane valley is followed in different portions of its course by the Kootenay, the Columbia, the Peace, and other great rivers of the Northwest. The trench of the Mid-Atlantic Ridge made headlines nationally when Maurice Ewing and Bruce Heezen announced that it not only continued along the entire length of the Mid-Atlantic Ridge but probably could be traced around

Figure 72. The principal relief features of the Atlantic Ocean. Trenches slightly exaggerated in scale.

the Cape of Good Hope into the Indian Ocean, with one branch extending up the Arabian Sea and another around Australia and up the eastern Pacific, a total of 40,000 miles. This claim has proven to be exaggerated. Some valleys along the South Atlantic ridge and in the Indian Ocean may be comparable to the trench in the North Atlantic, but these valleys are proving to be very discontinuous, and no central valley has been found in the ridges of the Pacific. The well-established portion of the trench is associated with many earthquakes and is in all probability a fault valley.

The recent studies of the Mid-Atlantic Ridge made by Bruce Heezen of Lamont and Dale Krause of Rhode Island University have indicated that there are a series of horizontal displacements along the ridge in the equatorial area. These are presumably due to faulting, and deep east-west troughs are found between the disconnected ridges. Heezen reported photographing ripple marks on the floor of one of these deep troughs, showing that strong currents may be flowing along them.

In addition to the Mid-Atlantic Ridge, a series of transverse ridges and rises extends out from the continents or from the Mid-Atlantic Ridge. To the north, the Iceland-Faeroe Rise and the Greenland-Iceland Rise form a shallow connection (2,400 feet) between Europe and Greenland, and this is continued west by the Baffin-Greenland Rise. To the south, a deeper connection, the Walvis Ridge, extends between Africa and the Mid-Atlantic Ridge and hence via the Bromley Plateau to South America. It was transverse ridges like these that led Bailey Willis, late of Stanford University, to suggest his idea of Isthmian links, which had the two sides of the Atlantic connected in the past by tongues of land like the Isthmus of Panama. These bridges were thought to have allowed animals to migrate across the oceans and thus account for the similarities of fossils both between North America and Europe and between South America and Africa. Proof of submerged bridges on the sea floor is still lacking.

The Atlantic basins and their deep-sea channels. On either side of the Mid-Atlantic Ridge there are long wide basins, mostly over 12,000 feet deep and partly deeper than 18,000 feet. In many places the floors of these basins are extraordinarily smooth, although gently sloping toward their deep points. In 1949, Lamont geologist Bruce Heezen discovered the first example of an elongate valley extending down the gentle slope of the Atlantic abyssal plain. Similar valleys have since been discovered by Heezen and by A. S. Laughton of the British National Institute of Oceanography. These valleys are generally referred to as *deep-sea channels.* These valleys have steep sides and flat floors that are from 100 to 600 feet lower than the surrounding ocean bottom. The valleys are from about three to five miles across. The best explored of these deep-sea channels appears to come down from Davis Strait and to extend around the Newfoundland Rise and hence down the Atlantic Basin toward the Nares Deep (Fig. 73). The valley appears to be continuous for at least several hundred miles. It has been explained by Maurice Ewing and his group as having been caused by excavations by turbidity currents, and they have at least one core from the bottom with sand, which suggests that a turbidity current had flowed along the valley. On the other hand, the fact that the deep-sea channels run roughly parallel to the trench of the Mid-Atlantic Ridge, which is certainly a fault valley, will convince many geologists that the channels are also due to faulting, perhaps a crack opening on the floor of the Atlantic. Turbidity currents could have flowed along such cracks, depositing the sands.

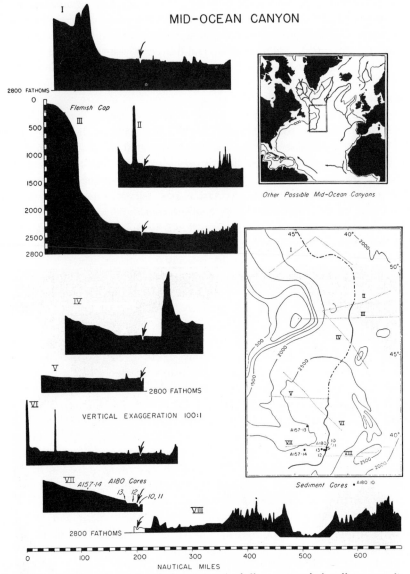

Figure 73. Showing the supposed directions of shallow entrenched valleys extending down each side of the Atlantic Ocean into the deep basins. Profiles from the fathograms are included. Data from Maurice Ewing and Bruce Heezen, Lamont Geological Observatory.

This explanation seems particularly likely for a valley that extends along the continental rise off Brazil, but a somewhat similar valley crossing the Biscay and Iberia abyssal plains off western Europe seems quite clearly related to turbidity currents coming from the continental slope off western France.

Floor of the Pacific

Trenches, ocean's greatest depths. Unlike the Atlantic, the floor of the Pacific is not distinguished so much by its ridges, although they are present, as by the line of trenches that extends virtually around the rim of this largest of oceans (Figs. 74, 77). In the Atlantic there are only four small trenches: two around the West Indies; one just east of the Mid-Atlantic Ridge near the equator, called the Romanche Deep; and one near the Antarctic. The trenches of the Pacific run continuously for as much as 2,500 miles (Peru-Chile Trench), and they have by far the greatest depths in the oceans. Up to a few years ago it

Figure 74. The principal relief features of the Pacific Ocean. Trenches slightly exaggerated in scale.

was thought that the Mindanao Trench had the greatest depth with its 34,428 feet, but now it is believed that the Challenger Deep of the Mariana Trench, discovered by the British, is the deepest with 35,616 feet[2] and that the Horizon Deep in the Tonga Trench, discovered by Scripps Institution, is second with 35,430 feet. Possibly the Ramapo Deep of the Japan Trench may also be in this category, although soundings taken on a recent Scripps Institution cruise fail to bear out earlier contentions and suggest that 33,000 feet is about the limit of depth for this trench. The Russians have come up with a candidate from the Kuril Trench that is a close contender, the Vityaz Deep given as 34,020 feet.

Some idea of what impressive holes in the earth's surface are constituted by the deep Pacific trenches can be seen from a comparison between Mt. Everest and the Tonga Trench (Fig. 75). The trenches located in the western Pacific

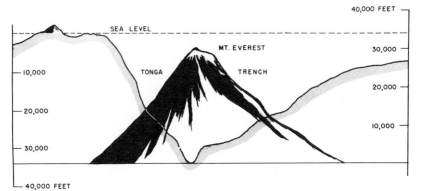

Figure 75. Comparison of Mt. Everest with the Tonga Trench. Both drawn to same scale (vertical exaggeration times 16). Courtesy of R. C. Winsett, Scripps Institution of Oceanography.

clearly constitute the greatest unevenness of the earth's surface, although the much shoaler Peru-Chile Trench with a maximum sounding of 25,284 feet has one wall that, including the west side of the Andes, rises about 42,000 feet.

The best-explored trench in the world, known as the Middle America Trench, extends south from the southern end of the Gulf of California almost to Panama. The soundings, made mostly by Scripps Institution, have been contoured by R. L. Fisher. He found that the floor was V-shaped in part and the rest had a flat surface several miles across. By timing reflections from dynamite charges set off under water, Fisher and George Shor, both of Scripps Institution, determined that the flat-floored portions are underlain by a thick mass

[2] Information from R. F. Fisher shows that the *Stranger* found depths of 35,700 feet, and this was confirmed by the dive of Jacques Piccard in the bathyscaph *Trieste* in January, 1960. These depths came from the same area where the *Vityaz* reported 36,000 feet in 1957.

of sediments, which are lacking where the floor becomes V-shaped. Toward land the wall of the trench is cut by several submarine canyons. As far as could be told, these canyons do not extend near the floor of the trench. The floor has a series of basins of quite variable depth, but none exceeding 22,200 feet. Some hills rising from the floor are presumably submarine volcanoes. An artist's concept of the trench at the north end of the Gulf of Tehuantepec is given in Figure 76.

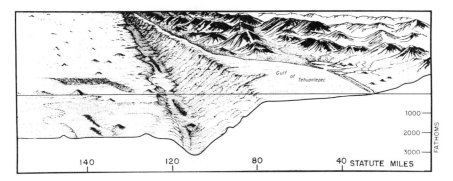

Figure 76. Block diagram, by R. C. Winsett of Scripps Institution of Oceanography, showing the character of the Middle America Trench as it extends along the west coast of Mexico. The mounds represent volcanoes on the sea floor.

Trenches related to earthquakes, volcanoes, and island arcs. The special relationship of the trenches to earthquakes, volcanoes, and other phenomena has been pointed out by Harry Hess of Princeton University, Beno Gutenberg and Charles Richter of California Institute of Technology, and others. There is little doubt that the trenches are due to crustal movements such as faulting. Earthquakes are more common along the trenches than in any other zone of equal area in the world. Virtually all of the tremors that occur along the trenches are of shallow crustal origin, whereas landward of all the trenches the earthquakes have origins at increasing depth until along a line about 200 miles landward of some of the trenches the earthquakes are deep focus, that is, they have depths of origin greater than 200 miles (Fig. 77). This relationship seems to exist for all trenches of the world. Furthermore, the trenches are missing wherever the earthquakes do not deepen in a landward direction. The assumption has been made that the pattern of earthquakes implies a great fault zone dipping under the continental margins (Fig. 77, inset).

Lines of volcanoes run parallel to many of the trenches, lying approximately on top of the zones with intermediate-focus earthquakes. Similarly, there are arc-shaped island chains like the Aleutians, the Kurils, and the Marianas, all occurring on the convex side of the arced trenches. Most of these islands are volcanic.

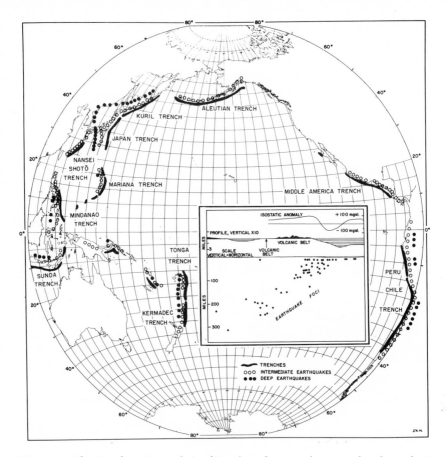

Figure 77. Showing the various relationships of sea-floor trenches to earthquakes, volcanic belts, and earth gravity (isostatic anomaly). The example shown in the inset is from Japan.

Measurement of the pull of gravity has shown still another unusual relationship in the deep trenches (Fig. 77, inset). Over virtually all trenches where measurements of gravity have been made from submarines, a great deficiency of mass has been found under the trenches.[3] It is as if there were a vacuum down there. Since such cannot be the case, it is assumed that there is a zone of down-buckled crust containing relatively lightweight rocks. It may be that the crust is held down against the force of gravity by pressure from the sides. This idea of Vening Meinesz has, however, been disputed by Maurice Ewing and others.

[3]Most of this work has been accomplished by Vening Meinesz of the Netherlands and J. Lamar Worzel and M. Talwani of Lamont Geological Observatory. It is now possible to make these measurements from surface ships.

The seamounts and guyots of the Pacific. There are seamounts and flat-topped guyots in the Atlantic, but they are not very numerous. In the Pacific, perhaps because of more exploration, seamounts have proven to be extraordinarily abundant and widespread. Guyots are more limited in extent and appear to occur along three lines (Fig. 78): one, the Emperor Seamounts south of Kamchatka; another in the Marcus-Necker Rise, west of the Hawaiian Islands; and the other in a zone extending from the Marianas to the Marshall Islands. In addition there are ten guyots all close together in the Gulf of Alaska that

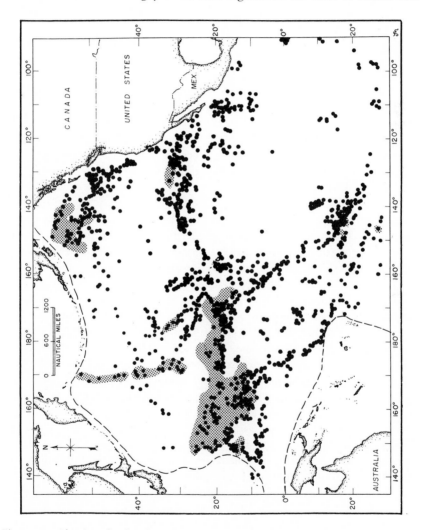

Figure 78. Showing the distribution of mountains rising 3,000 feet or more above the floor of the Pacific Ocean as islands, reefs, seamounts, and guyots. Most of those in the shaded area are guyots. From H. W. Menard, Scripps Institution of Oceanography.

may be in parallel lines. Several other guyots occur as isolated mountains. The flat summits are variable in depth, but those in the western Pacific are close to 4,800 feet and the ones in the Gulf of Alaska average about 3,000 feet. Combining islands, shallow banks and reefs, seamounts, and guyots, the Pacific is literally peppered with these mountains that rise thousands of feet above the deep-ocean floor (Fig. 78). Many of the blank places left on the combination map will no doubt be filled by future expeditions. Every expedition over uncharted waters brings in new submarine mountains, but new guyots are rarely discovered unless soundings are made along one of the rises where most of the previously known guyots have been discovered.

Explanation of seamounts and guyots. Mountains in the deep ocean do not create any mystery. Ever since Darwin's cruise of the *Beagle* the importance of vulcanism in the ocean has been appreciated. A volcano erupting on the sea floor can build a mountain in almost the same way as on the land surface. When the eruptions occur at great depth, however, they are not likely to have as much explosive activity as in shallow depth or above the surface because of the great pressure of the water. Also the lava cools faster on the sea floor, producing steeper slopes. Rocks collected from oceanic islands or dredged from seamounts have proven to be almost 100 per cent volcanic, mostly the dark lavas referred to collectively as basalts.

The guyots are more of a puzzle. Some volcanoes have a flat top where an even-crested crater has been filled with lava, but most craters have an uneven summit, so they could not readily form the plateau top. Furthermore, the terraces on the sides of some seamounts cannot be explained by vulcanism. It seems far more likely that flat-topped seamounts have once been at or near sea level. The waves can cut off the summit of a newly erupted volcano in a very short period because so much of the summit of the peak consists of crumbling ash and volcanic slag deposits, which are rapidly attacked by the great mid-ocean waves. The fact that the depth of a guyot summit is slightly greater on the margins suggests wave erosion. In tropical areas, further flattening may be brought about by growth of coral reefs (see Chap. X). Most coral reefs, however, have rimmed margins. The sinking of a flat-topped coral bank would probably be accompanied by upgrowth along the margin even if the sinking were too rapid to allow the reef to grow up at the same pace. Therefore, if sunken reefs do account for many of the guyots, we should find such rims. They have not been found to date. Another argument against sunken reefs is that the margins of reefs are usually much steeper than the upper slopes of guyots. On the other hand, coral did grow on at least two of the guyots, as was shown when the Scripps Institution and Navy Electronics Laboratory expeditions dredged ancient corals from them. Also it seems very likely that rows of atoll islands like the Marshalls and the Gilberts have grown up on the top of what would have become guyots had it not been for the ability of the coral to grow up as the guyots sank.

Assuming that most of the guyots are the result of wave planation and sub-mergence of the beveled platform, how do we explain the submergence? There are two prevalent ideas, one that the sea level rose in relation to a subsiding ocean floor along lines of crustal weakness, and the other that the guyots have sunk. The former was suggested by Harry Hess, who first called attention to the guyots. He thought that they were islands beveled during the pre-Cambrian, that is, a billion years or more in the past, and that the sedimentation in the oceans since that time has been great enough to raise the sea level by displacement of the ocean water by sediment and at the same time to bend down the earth's crust. This idea has run into two difficulties. One, the age of beveling of at least some of the guyots has now been determined by study-ing fossils collected from their surfaces. Corals from a guyot in the group west of the Hawaiian Islands have been identified as Cretaceous by E. L. Hamilton of the Navy Electronics Laboratory, and shallow-water foraminifera from a guyot west of California have been identified as Miocene by several California micropaleontologists. Another guyot off Newfoundland also proved to be Mio-cene. In both cases the surfaces must have been at sea level a long time after the pre-Cambrian. Furthermore, as will be shown in Chapter IX, the sediments of the deep ocean do not appear to be nearly thick enough for the subsidence that Hess had conceived.

Roger Revelle of Scripps Institution has suggested that the sea-level rise has been more recent, having started at the end of the Cretaceous as the result of great lava extrusions that brought up enormous quantities of gas, which became condensed as water. The ocean floor sank due to removal of the gas, and the sea level rose because of its addition as water to the ocean. As an argument for this history, Revelle has referred to the enormous quantity of limestone deposition that has occurred since the Cretaceous. Probably that was the time when the globigerina oozes and other calcareous deposits started to blanket much of the ocean floor. The rate of consumption of carbon dioxide by the organisms to allow this rapid deposition of calcium carbonate seems to require a tremendous amount of vulcanism because the atmosphere does not have nearly enough carbon dioxide.

The idea of regional subsidence that is favored by Menard seems to have some advantage in explaining the guyots because they are so linear in their dis-tribution. Seamounts are widespread through the Pacific. Assuming that there was a geological revolution in the Cretaceous, one would expect that the islands with flat tops that were submerged as guyots at the time would have a random distribution more like that of the seamounts. It does not seem likely that in the vast areas without guyots all of the seamounts have been formed since the Cretaceous. Perhaps the answer is a combination of the two ideas. In any case the guyots do not have the same levels, although guyots in the same area have similar depths. Those in the Gulf of Alaska show possible evidence of having been downwarped because a seamount in the Aleutian Trench has a far deeper summit than those to the south (Fig. 79). Another evidence of downwarping of the ocean floor was found, by E. L. Hamilton of the Navy Electronics Laboratory and R. S. Dietz, in the depression that surrounds the Hawaiian Islands and indicates that this mass has constituted a burden on the earth's crust and has bent it down.

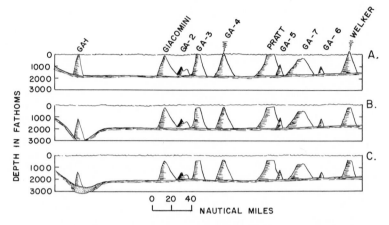

Figure 79. Suggested order of development of seamounts and guyots in the Gulf of Alaska. Sec. A: The seamount on the far left is about the same level as the others. Sec. B: Because of the development of the Aleutian Trench, the same seamount has been carried well below the surface. Sec. C: The same seamount is still lower and is partly buried by sediments in the bottom of the trench. From H. W. Menard, Scripps Institution of Oceanography. Recent studies of the Navy Electronics Laboratory show the seamount in the trench is not a guyot.

Ridges or fracture zones off western North America. One of the most surprising oceanographic discoveries of recent years has come from the work of H. W. Menard on the soundings obtained in the area off the west coast of North America. Despite the fact that the bathymetric charts of the oceans published by the International Hydrographic Bureau at Monaco show nothing of interest in the area,[4] Menard found that "there are four great bands of unusually irregular topography" extending out from 1,600 to 3,300 miles into the Pacific and having an average width of 60 miles (Fig. 80). He called these bands "fracture zones" because of the many fault scarps, narrow ridges, and narrow fault troughs that extend in some cases as much as 1,000 miles along these zones. Many seamounts occur and there are a few volcanic islands. At least three of the zones connect with the continent. The northernmost, the

[4] These charts are badly out of date, and no adequate organization has existed to keep them abreast of the times. The world chart issued in 1961 by the Oceanographic Office of the U.S. Navy and the Soviet Union bathymetric chart of 1965 give a far better picture of the relief of the oceans and should be used for teaching purposes to replace those of Monaco.

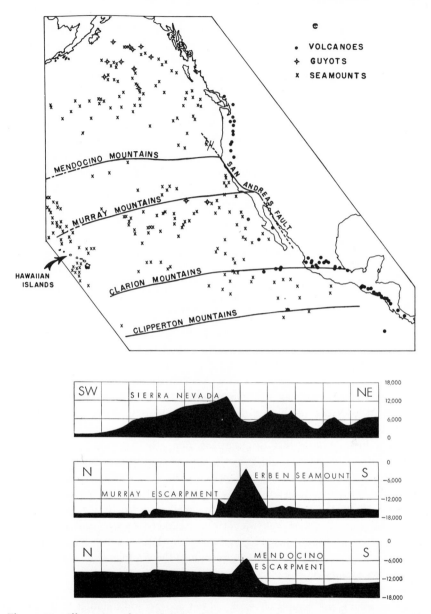

Figure 80. Illustrating the profusion of seamounts and guyots in the eastern North Pacific Ocean and the mountain ranges that extend out from the west coast of North America. Courtesy of H. W. Menard, Scripps Institution of Oceanography. Note the comparison of a profile of the Sierra Nevadas with two submarine profiles. The vertical scale is in feet. The horizontal intervals are ten miles. Profiles reproduced through courtesy of Scientific American.

Mendocino fracture zone, lies directly off the Gorda escarpment (Fig. 52), suggesting possible connection with the San Andreas Fault, but curiously enough the Mendocino sea scarp, which is a cliff a mile high in places, slopes to the south, whereas the Gorda escarpment slopes north. There is a narrow ridge on the north side of the sea scarp. Together, because of their continuity and straightness, they provide a remarkably fine means of determining latitude at great distances from the land. To the south of this fracture zone the water is about 3,000 feet deeper than to the north.

The investigation of the Mendocino fracture zone by Victor Vaquier and his colleagues at Scripps Institution has shown that the variation in magnetic intensities in east-west line, which were run respectively north and south of the great escarpment, can be matched only by a horizontal shift along the escarpment of as much as 600 miles. This suggests that the earth's crust north of the escarpment has moved 600 miles to the west in relation to the crust on the south side. Such a displacement agrees with the shift of the 2,200-fathom contour shown in Figure 74. It will be seen that this displacement is in the opposite direction from that found along the Gorda Escarpment (Fig. 52) that lies directly landward of the Mendocino fault. This is one of the greatest enigmas ever discovered by science.

The Murray fracture zone appears to be a continuation of the east-west ranges of southern California that extend just north of Los Angeles and Santa Barbara. Here, also, the San Andreas Fault is deflected rather sharply to the west (see p. 106) but resumes its northwest trend after a short distance. The Murray fracture zone has less continuous sea scarps, but they slope mostly northward with the result that the region between the Murray fracture zone and the Mendocino fracture zone is deeper than that on either side. Magnetic intensity lines indicate that displacement along the Murray Escarpment zone has a shift of 96 miles in the opposite direction to that along the Mendocino Escarpment, according to available information. The Murray fracture zone has numerous seamounts, and the only two guyots off California are located along this line. The Clarion fracture zone is less well explored. It appears to connect with the great east-west volcanic belt of southern Mexico and has volcanic islands along its length including San Benedicto Island, which has had a recent eruption. It is interesting that 150 years ago the explorer and scientist Alexander von Humboldt observed this east-west trend in Mexico and thought that it might continue into the islands reported west of that area. The Clipperton fracture zone is not completely explored but appears to be similar to the others. It was named after Clipperton Island, which occurs along its length.

Archipelagic aprons, sea fans and left-hooked deep-sea channels. The work of Menard has also indicated that there are many smoothly sloping areas in the Pacific. Around a number of the island masses he found apron-like slopes extending to the deep-sea floor. These archipelagic aprons, as he called them, appear to be largely the result of lava flows built out from the islands or of

fissures erupting on the submerged sides of the islands. In places the lava surfaces have been covered with recent sediment, but elsewhere the lava may be
exposed directly on the slope.

Off many of the large rivers of the west coast there appear to be gently
sloping coalescent fans cut by deep-sea channels that can be traced down toward the deep part of the basins (Fig. 81). These are similar to the deep-sea
fans bordering the Atlantic continental slopes. The fans, as far as they have
been sampled, appear to have turbidity-current sediments. Presumably also the
deep-sea channels are related to turbidity currents (see p. 23). The most
curious thing about the channels is that all of them hook to the left, so that
they follow the southern or eastern portion of the fan. The same is true of the
much shoaler channels that form an extension of the small submarine canyons
of southern California. Menard has suggested that this left hook of the chan-

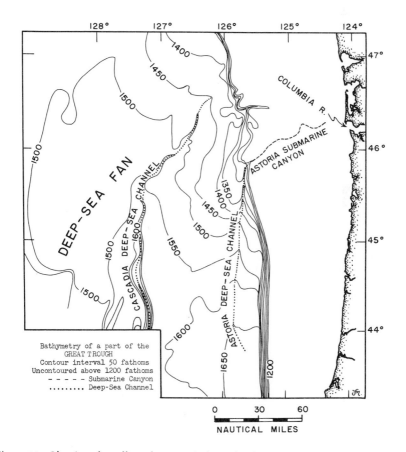

*Figure 81. Showing the valleys that extend along the deep-sea fans outside the Columbia
River. Modified from H. W. Menard, Scripps Institution of Oceanography.*

nels is the result of earth rotation, which causes currents to turn to the right in the Northern Hemisphere (e.g., the Gulf Stream). The right turn of the turbidity currents on the sea floor causes more deposition on the north side of the fan, so that eventually the south side receives the channel because it is the lowest point. The idea is intriguing but certainly needs more evidence, particularly from areas other than the west coast of the United States. We should look for the channels along other shores and in the Southern Hemisphere, off the Congo Canyon for example, because the rotation of the earth has an opposite effect south of the equator.[5]

Floor of the Indian Ocean

The numerous Indian Ocean expeditions conducted by many nations in the 1960's have made this ocean as well known as the Atlantic and the Pacific. The Bruce Heezen and Marie Tharp physiographic diagram, published by the Geological Society of America in 1964, provides a vivid picture of the complicated relief of the floor of this ocean (see Fig. 82). The most striking feature is the great ridge that has two converging arms extending up into the Arabian Sea like a great inverted "Y." This is thought by Heezen and others to represent a continuation eastward of the Mid-Atlantic Ridge, which bends to the east and extends past the south side of Africa at a distance of 900 miles. Like the Mid-Atlantic Ridge, the great ridges of the Indian Ocean show indications of horizontal displacement by faulting (Fig. 82). These faults run north and south, unlike the displacements of the Mid-Atlantic Ridge that run east and west. Another north-south feature is the recently discovered ridge called *Ninetyeast* (because it lies close to that longitude). It extends up into the southern portion of the Bay of Bengal.

The only significant trench in the Indian Ocean, found south of Sumatra and Java, follows the arc made by Indonesian islands. Its greatest depth is 24,390 feet. It appears to be the easternmost offshoot of a series of arc-shaped trenches found in the western Pacific, but is not nearly as deep as most of the others.

Two curious features are found off the Ganges and Indus Rivers, respectively. Here, we have broad, gently sloping fans (called *cones* by Heezen and Tharp) that extend seaward for almost 1,000 miles, and, except for shallow deep-sea channels that branch and wind across the plains, the floor of the ocean is very smooth. Apparently, an enormous quantity of sediment has been carried south of the mouth of these two rivers, which are the largest of any

---

[5] The recent soundings reported by Bruce Heezen of Lamont show that the extensive valley system crossing the fan off the Congo Canyon hooks to the left in spite of being in the Southern Hemisphere.

entering the Indian Ocean. This sediment has filled up the floor of what may originally have been an irregular basin and made these smooth plains.

Generalizations Concerning Deep Oceans

The examples of submarine topographic types that have preceded have come entirely from the two largest and best-known ocean basins, but the other oceans are all almost as deep, that is, largely more than 12,000 feet. Preliminary studies

Figure 82. Diagrammatic sketch showing the relief provinces in the Indian Ocean. From descriptive sheet accompanying "Physiographic Diagram of the Indian Ocean," by Bruce C. Heezen and Marie Tharp, published by Geological Society of America.

of the Indian Ocean indicate that it has many of the same features, including a large ridge along the center of the Arabian Sea, and many transverse ridges. The Arctic Ocean also has at least one main ridge, which Soviet soundings have shown extends from the New Siberian Islands to Ellesmere Island, west of Greenland. The deep landlocked seas, such as the Gulf of Mexico, the Caribbean, the Mediterranean, and the Black Sea, are all comparable to ocean basins in their depth, since they exceed 12,000 feet in large areas. The Gulf of Mexico and the Black Sea are comparatively flat floored, whereas the Mediterranean and Caribbean have numerous ridges and basins.

The average depths of the three largest oceans are as follows:

	Depth, excluding adjacent seas	Depth, including adjacent seas
Pacific Ocean	14,040 feet	13,200 feet
Atlantic Ocean	12,900 feet	10,920 feet
Indian Ocean	13,080 feet	12,840 feet

The most interesting thing about these averages is that the Pacific with all its great trenches is not much deeper than the Atlantic and Indian oceans, which have virtually no trenches. In fact all of the oceanic basins, including the landlocked seas, have about the same depth and thus appear to have the same reason for being so much deeper than the continents. The seismographic exploration described in Chapter IX gives what appear to be good reasons for the existence of these basins.

Sediments of the Deep Ocean

Coring apparatus. Coring in the deep ocean began with the *Challenger* Expedition of 1872-76, although cores at that time were only about one foot in length. In the present century only a few deep-ocean cores had been obtained up to the end of World War II. The German *Meteor* Expedition was responsible for a large proportion of these. Scientists on the Woods Hole Institution's *Atlantis* had begun to obtain many cores of up to about ten feet in length in various parts of the Atlantic. The *Snellius* Expedition of 1929-30 had cored in the East Indies, which includes many deep basins, and Charles Piggot of the National Research Council had obtained cores of up to ten feet in length, using a gun to shoot the corers into the bottom. This gun, however, had too many near misses in shooting holes into the sides of vessels, so it has been abandoned.

During World War II Börje Kullenberg, who now heads the Oceanographic Institute at Göteborg, Sweden, made use of the piston principle in developing an instrument that was so simple that many oceanographers have been wondering why they had not thought of the idea previously (Fig. 83). The piston

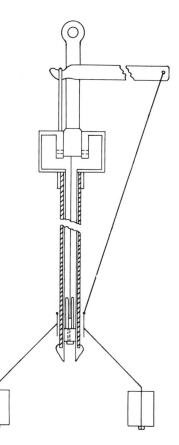

Figure 83. Illustrating the principle of the Kullenberg piston corer. The trip weights are shown on the side, and the piston in the center of the core barrel. (Gaps left to conserve space.)

corer has a trip weight that is suspended below the device. When the trip weight reaches bottom, it releases the core barrel, which falls free, while the piston that is inside the core barrel descends at a much slower rate. The brakes of the winch on the deck of the vessel are set at the time when a tension gauge gives a sudden jump, indicating that a great weight has been released from the line.[6] This prevents coiling up of wire on the bottom. The core barrel falls free for its entire length, while the piston remains immobile. This piston action greatly reduces the effect of friction that would normally result from the core being pushed into the core barrel. This is a principle that is used by many types

[6] The hitting of the bottom is also indicated by what is known as a *pinger*. This device, attached to the corer, sends out continuous sound signals and shows when the corer is approaching the bottom by the coalescing of two lines on a fathogram. When the separation is just right to indicate that the device has hit bottom, the descent of the corer is stopped.

of soil samplers on land, but in the deep ocean the tremendous hydrostatic pressure, which is built up at the rate of one additional pound per square inch for each additional two feet of depth, acts as a differential force to drive the core barrel into the bottom. This device is one of many examples of the large contributions that the Scandinavians have made to the science of oceanography.

Glacial stages and deep-sea cores. If you had consulted almost any geologist 30 years ago, he would probably have told you that long cores in the deep ocean would be an expensive way of finding out more about the same material you could get in a short core. The ocean must have been an ocean for a long time, so that the continuous rain of small marine animals, the dust carried out in the atmosphere and settling slowly to the great depths, and the sediment carried out from rivers or by wave action were thought to be a small but fairly constant source. Because other sediments were contributed so slowly, meteorites were said to be important in the deep-sea sediments. With these sources there seemed to be no particular reason for believing that the sediment on the sea floor should be stratified or change in character from time to time. As has so often been the case in the past, this idea, built up without much factual background, proved to be erroneous. Actually the long deep-sea cores have shown a remarkable series of alternating layers.

One of the first indications that all was not well with the old idea of slow accumulation of the same kind of material for millions of years was brought to light by the Piggot corers shot into the ocean bottom along a line traversing the northern Atlantic Ocean. The studies of the cores showed M. N. Bramlette and W. H. Bradley of the United States Geological Survey that the deposits of the great ice age were represented in the sediments lying at depths about one foot below the bottom. It stands to reason that during the times when the great glaciers covered northern North America, Greenland, and a considerable part of Europe great numbers of icebergs broke off from the floating margin and drifted out across the North Atlantic. These icebergs carried great quantities of stones, which were dropped as the ice melted. Since other types of deposition in the deep ocean were relatively slow, just as they are now, the stones are well represented in the deposits of these glacial stages. The Piggot cores across the North Atlantic contained layers with stones and gravel interbedded with other layers that represented the relatively warm episodes sandwiched in between the times when the earth was refrigerated.

It has been possible to connect some of the stages of the great ice age from core to core across the Piggot Atlantic traverse. Independent evidence of alternating cold and temperate climate came from a study of the character of the foraminifera in the various layers. Some of these unicellular animals that drift around near the surface of the ocean are very sensitive to temperature. Experts have learned to recognize the types that are restricted to various temperature conditions (Fig. 84). As the temperatures changed with the development of the ice age, the foraminifera now living in the Arctic seas moved into lower

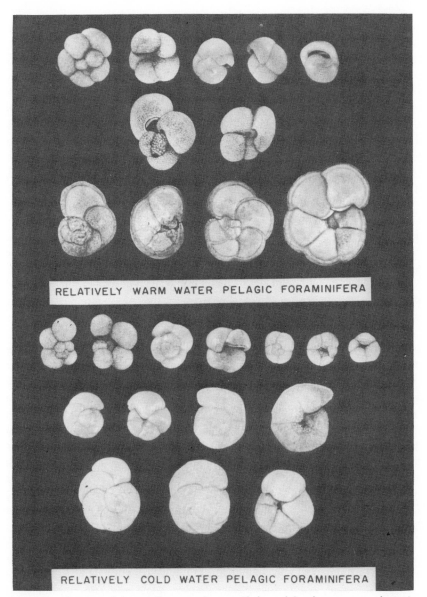

Figure 84. A group of foraminifera greatly magnified, used by the experts to determine whether the foraminifera were living in relatively warm or relatively cold water. Courtesy of F. B Phleger, Scripps Institution of Oceanography.

latitudes. Thus the foraminifera at a little depth below the top of the cores are of a different character from those in the surface sediments.

The most complete study of these glacial and postglacial sediment layers in the deep sea has been made by Lamont Observatory's David Ericson. His study of the foraminifera, combined with carbon-14 age determinations by W. S. Broecker and J. L. Kulp, both also of Lamont Observatory, and oxygen-18 temperature measurements[7] by Cesare Emiliani of the University of Miami showed that the waters of the Atlantic warmed up about 11,000 years ago. This gives a date in the midst of the retreat of the great ice sheets that accords with other lines of evidence. The extent of the last glacial stage, which Ericson believes is about 50,000 years, is more open to controversy, others considering it much shorter.

Deep-sea sands. Geologists were not particularly disturbed by the discoveries that the great glacial period was represented by a change in sediment in the higher latitudes, but considerable surprise and some disbelief came from the later discovery that sand beds, some of them as clean as those found on the beaches and in the shallow water near shore, are well represented on portions of the deep-sea floor. The surprise became even greater when it was reported that these sand layers contain shells of foraminifera that are now living only on the bottom of shallow seas.

At first there was some thought that these deep-sea sand layers might represent great submergences, or, in other words, that the deep-sea floor had once been near the surface, where waves and currents could be expected to produce the coarse sediments. However, with the advent of the piston coring devices and the obtaining of long cores it was found that the sand layers alternate in many places with typical deep-sea deposits. The distribution of sand layers in the North Atlantic as interpreted from the Lamont coring operations has been quite adequately mapped by David Ericson (Fig. 85). The indications are that these sands extend out from various parts of the coast in great sea fans. They are obviously one of the manifestations of the great transporting power of turbidity currents. On the other hand, the fact that only 134 out of 550 Lamont deep-sea cores have sand layers suggests that a large part of this ocean may not be receiving the coarser turbidity-current deposits.

The other more common types of deep-sea deposits include deep-sea oozes, brown clays, and terrigenous muds. The distribution of these deposits as far as it is known is indicated in Figure 86.

Deep-sea oozes. The soft squashy sediments known as oozes are found in general in the shoaler portions of the deep-sea floor. The oozes consist of the minute shells or skeletons of the low types of animals and plants that drift in the surface and near-surface waters and are known as *plankton*. Of these

---

[7] A method formulated by Harold Urey of the University of California, San Diego, that uses the ratios between oxygen-18 and oxygen-16 isotopes obtained from shells to determine the temperature of the water in which the shells lived.

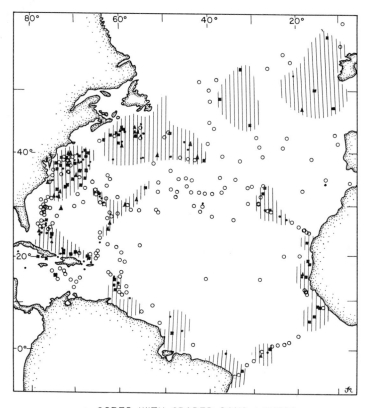

CORES WITH GRADED SAND LAYERS
CORES WITH SAND LAYERS NOT GRADED
CORES WITH SILT LAYERS
CORES WITHOUT SAND OR SILT LAYERS

Figure 85. Location of cores with sand and silt layers (shaded areas) found on the deep floor of the Atlantic Ocean by Lamont Geological Observatory. Modified from D. B. Ericson et al. in Crust of the Earth, Geol. Soc. of Amer., 1955.

the foraminifera are the most abundant and because the common forms are *Globigerina* (Fig. 84), the deposit in which they dominate is referred to as *Globigerina ooze*. This type of ooze is found widely in the Atlantic and over rather extensive areas in the South Pacific.

Diatom ooze is dominated by the remains of a siliceous plant that is very common in the upper waters, at times developing almost a soupy appearance at the surface. The deposit is found in a large belt around the Antarctic and in another extensive area northeast of the Japanese Islands. In both of these areas oceanic waters of intermediate depth, which are rich in nutrients such as phosphates and nitrates, are brought up to the surface and provide the food for the plants. Large numbers of animals live on these plants, but not very

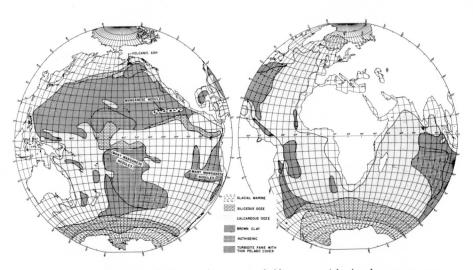

Figure 86. Distribution of deep-sea sediments. Probably zones with abundant manganese nodules cover more territory than is indicated and turbidite layers are interbedded with normal pelagic deposits in more areas. All boundaries are only approximate. From Submarine Geology, *2nd edition, Harper & Row, Publishers.*

much of the animal remains gets to the bottom because they are more soluble than the silica in the plants. In a few areas around the equatorial Pacific the siliceous animals known as Radiolaria are predominant in the deposits and constitute another variety of ooze.

Brown clays. A sediment that is predominantly clay has been found over most of the deeper portions of the ocean basins. The older name *red clay* for this material is unfortunate, since most of these deep clays are brown or buff colored. It happened that the first samples that were obtained in the South Atlantic had a red color, so the name was established for all similar clays. The low content of calcium carbonate in the clays was at first puzzling in view of the enormous masses of free-floating organisms in the oceanic waters that must be contributing their skeletons to the ocean floor. Then it was realized that solution was sufficient in the areas because the cold carbonated waters coming from the Antarctic and moving along the deep basins remove the calcium carbonate as it sinks or dissolve it out of the slowly accumulating sediments. According to Robert Rex of California Research Corporation the clay comes largely from atmospheric dust, from fine-grained land-derived deposits that have circulated for great distances in the major ocean-current systems, to some extent from meteorites, and to a minor degree from volcanic dust. Some of the deposits that were formerly called *red clay* are now found to consist primarily of phillipsite, a mineral which has been shown by Gustaf Arrhenius of Scripps Institution to have been crystallized out of the sea water and deposited widely over the deep portions of the Pacific Ocean at places that are far from good sources of clay from the lands.

Terrigenous muds. The deposits that flank the continents and constitute the principal sediment of the deep-sea fans are referred to as *muds.* They differ from the brown clay in having a considerable quantity of silt, and as we have learned recently, they are often interlayered with sands. These muds are predominantly land derived, probably having been transported largely by turbidity currents. The colors of the muds show a relationship to the adjacent land masses. If large rivers enter nearby, the muds are apt to be red, representing the oxidized conditions of the river deposits. Outside dry coasts the muds are usually green. Black muds are found where stagnant basins occur along portions of the coast and where a large amount of plant and animal life is contributed. This organic debris may form future oil deposits. White muds occur on some of the steep slopes around coral islands and consist of the pulverized fragments broken from corals and other calcareous organisms of the reefs. This fine material is carried down the slopes, partly by landslides.

Nondepositional zones on the deep-sea floor. All dredgings that have been made on the seamounts or on smaller elevations of the deep-sea floor have shown that deposition of recent material, if present at all, occurs in very limited quantities and extent. This was brought home to us in 1938 when we tried to get a core at a depth of 12,000 feet on top of a gentle rise coming up from a surrounding sea floor of 14,400 feet. After working in a rough sea for hours to disentangle the last 2,400 feet of wire, we found that our core barrel had been badly smashed by hitting a rock covered with manganese. Small fragments were caught in the bent tube. One of the guyots off California that we photographed showed scattered *Globigerina* ooze deposited in cracks of the rock surface (Fig. 87). Virtually all of the Lamont photographs on Atlantic seamounts show rock. Much of the seamount rock is coated with manganese, and manganese nodules are common on the ocean floor, in general indicating very slow deposition.

Perhaps more surprising is the existence of extensive zones of nondeposition on relatively flat floors of the deep Pacific. These have been established by the work of W. R. Riedel of Scripps Institution on Radiolaria obtained first on the Swedish Deep Sea Expedition. Out of fifteen piston cores taken in runs from Tahiti to Hawaii and then back to the Ellice Islands (near Fiji), Riedel found that Tertiary Radiolaria occurred at or near the surface in eleven of the cores; in many localities the older Radiolaria are mixed with recent sediment. More recently, Riedel has examined hundreds of Scripps Institution cores from the central Pacific and has found zones extending roughly east and west that show increasing age of the sediments going north from the low latitude, centering a little north of the equator, with high recent calcareous deposition. The oldest sediments are of Oligocene age (about 35 million years old) and are scarcely covered at all by more recent sediments. This shows that deposition

Figure 87. Photograph of the rock surface of a guyot off the west coast of the United States. The white indicates Globigerina ooze deposit in the cracks, and the spiked animal is an echinoid. Navy Electronics photo A. J. Carsola and R. S. Dietz.

in this tropical central Pacific has been amazingly slow or locally absent and that the older material is being redistributed on the ocean floor. In some localities currents of the deep-sea floor appear to be capable of removing much or all of the finer sediment that is now sifting to the bottom, allowing only some of the coarser material, such as foraminifera, to accumulate.

Ever since the *Challenger* Expedition it has been known that manganese nodules exist on the deep-ocean floor. Recent oceanographic expenditions, particularly those in the Pacific, have indicated that manganese may be so abundant on parts of the ocean floor that some day it may be possible to recover it for commercial uses by dredging, partly for the cobalt and nickel that the nodules contain. Photographs often show large clusters of these nodules protruding from the sea floor (Fig. 11). Some of them are up to three feet in maximum diameter. The nodules have been found as thick coatings on a large proportion of the rocks dredged from seamounts both in the Atlantic and in the Pacific. It is rare that any manganese is found in cores from the ocean bottom except at the top of the cores. These nodules raise a very curious problem. Why have they formed in relatively recent times but not at more remote periods? As yet there is not even enough information to be absolutely sure that this distinction exists, let alone to explain the difference.

Another mystery not yet explained is why the fossils found to date on the deep-sea floor and on seamounts are not older than the Cretaceous. This may indicate that the oceans are not very old. The drilling (called *Mohole*), into the ocean floor, planned originally to go down through the earth's crust to the mantle, has been sidetracked temporarily (1966) by the U. S. Congress. A less costly drilling project (known as *Joides*) was funded by National Science Foundation and will be undertaken cooperatively by Woods Hole, Lamont, Scripps, and Miami laboratories. This will drill to about 3,000 feet below the ocean bottom in many places and may provide information on the age of the ocean, because the drilling probably will pass through the various deposits formed on the ocean bottom ever since the basins were first developed.

CHAPTER IX

UNDER THE OCEAN BOTTOM

At first thought it might be supposed that a discussion of what is under the ocean bottom belongs in the realm of science fiction. Actually the information on the subject gathered in the past twenty years has aroused more interest among scientists than almost any other phase of ocean study. Sound waves, used so successfully in learning about the configuration of the sea bottom, have been used also as an important means of unlocking the secrets of the earth's crust under the ocean. The results, which are still somewhat speculative, are permitting geologists and geophysicists to explain many things about the nature of deformation of the crust and to interpret on a much better basis the history of the past. So much interest has been aroused by these studies that plans have been made for putting a drilling down for many miles into the deep-ocean bottom to check the new interpretations. This Jules Vernean scheme would have been greeted by scientists with derision a few years ago, but now that we are preparing to visit the moon and are drilling deep oil wells in the shallow part of the ocean floor, it does not seem unreasonable to extend the drillings deep into the deep-ocean floor. Furthermore, the knowledge gained from an oceanic boring may have some real importance to military science.

Methods of Probing by Sound

Early reflection methods. Almost as soon as echo soundings began to be recorded, fathograms that indicated a reflecting surface underneath the ocean bottom were observed locally (Fig. 88). The first method that succeeded in penetrating deeply into the ocean floor, bringing back substantial information about the deep sub-bottom layers, made use of high intensity sound waves from surface explosions. W. Weibull, of the Oceanografiska Institutionen of Göteborg, Sweden, first employed this method when he accompanied Hans Pettersson on the *Albatross* during the epoch-making Swedish voyage around the world in 1947-48.[1] The method was also used later by American oceanographic institutions and provided some of the earlier estimates of the total thickness of unconsolidated sediments on the deep ocean floors. It was not satisfactory, however, for determining thin surface layers and has now been superceded.

In the late 1940's, Woods Hole and Lamont began using low frequency sound waves of the order of a few hundred cycles to study the deep layers.

[1] Hans Pettersson, *Westward Ho with the Albatross* (New York: E. P. Dutton & Co. 1953).

These penetrated the bottom to a depth of several hundred feet and were more selective than the Weibull method. They were also based on explosive energy and thus could not show much about sub-bottom structures because of the long interval between setting of explosives.

Continuous reflection profiles. In the early 1950's, several groups, including the laboratory of Socony Mobile Oil Company, the U. S. Geological Survey, and Woods Hole Oceanographic Institution, began experimenting with electro-acoustic transducers, using low frequency pulses generated by electric sparks sent off in rapid succession from a device towed astern of a vessel that was underway at slow speeds. Using a recording tape, it became possible to obtain a continuous profile showing the structures of the layers under the bottom along the course followed by the ship. The method has been greatly improved as the result of modifications made by W. B. Huckabay of Dallas, J. B. Hersey of Woods Hole, and many others. It is now generally referred to as continuous reflection profiling. The methods are referred to also as *Arcer, Boomer, Sparker,* and *Air Gun,* all but the last using a large electric spark for the sound impulse. The Air Gun releases a compressed air bubble.

Many thousands of miles of profiles have now been completed across the principal oceans. These profiles have been run primarily by such investigators as John Ewing of Lamont, George Shor of Scripps Institution, and J. B. Hersey of Woods Hole for the deep oceans, and by J. R. Curray of Scripps and D. G. Moore of Navy Electronics Laboratory for the continental slopes.

The records from this work show structural profiles that, except for a considerable vertical exaggeration, are often vastly superior to the cross sections made by land geologists in their studies of mountain ranges. The profiles penetrate the bottom as much as 3,000 feet under favorable conditions. Examples shown in Figure 89 serve to illustrate the advantage of the method which acts as a sort of X-ray photograph of the layers underlying the bottom. It is now possible to proceed at as much as 11 knots, a good cruising speed, and obtain these records while crossing unexplored portions of deep oceans.

Refraction waves from explosives. In the early twenties geologists and geophysicists began using explosive waves to determine the existence of buried rock structures in the oil fields. This work was largely responsible for changing a threatened oil famine into undreamed-of wealth for the petroleum industry. Dynamite is exploded in shallow holes, causing artificial earthquakes, and the times of arrival of waves are measured on a series of small portable seismographs set up at various distances from the shot point. Waves traveling through the earth arrive at the stations along a variety of paths, and the times of arrival depend on the speed at which the waves travel through the various

Figure 88 (opposite). Showing the use of the Sonoprobe to determine the bedrock under the sediment and the dip of the underlying formations. The second echo is caused by a second return of the sound from the bottom after echoing from the surface. Courtesy of D. G. Moore and the Navy Electronics Laboratory.

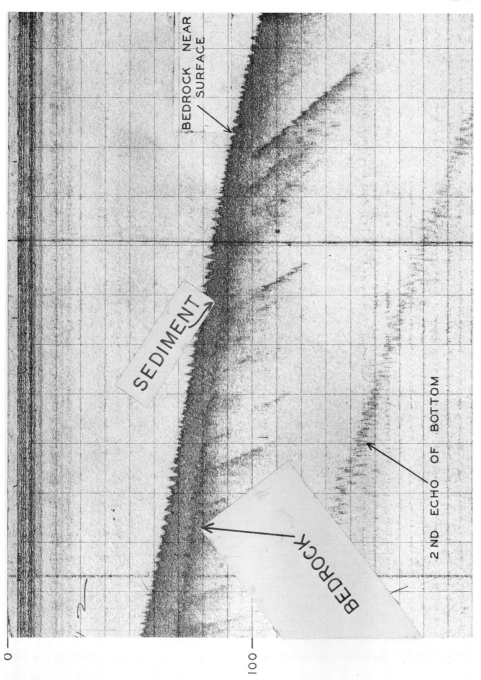

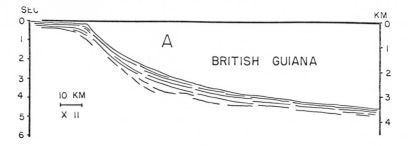

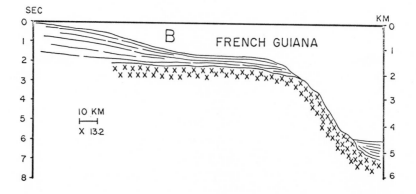

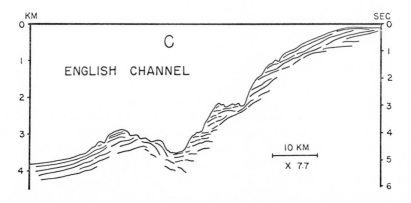

Figure 89. Continuous reflection profiles showing the structure under the continental slopes. Vertical exaggeration of each profile indicated. A. Shows building up of the outer shelf and slope by sedimentation. B. Illustrates outcropping of old rock on the continental slope with no depositional cover. C. Shows forward building of the continental slope followed by erosion of the continental slope forming submarine canyons and slumping. These records, furnished through the courtesy of J. R. Curray of Scripps Institution and D. G. Moore of Navy Electronics Laboratory, were taken on ships of Scripps Institution and the British National Institute of Oceanography.

layers encountered en route. When the waves traveling down from the source encounter a hard layer in which vibrations travel more rapidly, they are refracted toward the horizontal and in many cases follow the contact of the hard layer, sending waves back diagonally to the surface all along the way (Fig. 90). Because of the high speed of travel possible along these deep layers, the waves traveling along them may reach the seismograph before the more slowly traveling waves in the surface layers. In cases where the waves traveling in the surface layer have arrived first, these surface waves may have been sufficiently deadened by the time of the second arrival that the next waves are detected in the seismograph record.

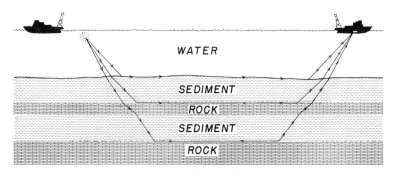

Figure 90. Showing the use of sound refraction and multiple echoes to determine the thickness and the nature of the rock and sediment under the ocean bottom. Note that when the sound travels from rock down into underlying sediment, the ray is bent downward so that no return comes from the surface between the two.

The use of these refraction waves for studying the ocean floor was begun in 1937 on Woods Hole Oceanographic Institution vessels by Maurice Ewing and his associates Allyn C. Vine of Woods Hole, J. Lamar Worzel of Lamont Geological Observatory, and George P. Woollard of the University of Hawaii. Results before World War II were spectacular but rather deceptive. Directly after the war, work began again and has centered at Lamont Geological Observatory under Maurice Ewing, at Scripps Institution under Russell W. Raitt, at Woods Hole under J. Brackett Hersey, and at Cambridge University under Sir Edward C. Bullard and the late Maurice N. Hill.

At first, bombs were lowered to the ocean bottom before exploding, and the vibration waves were received from a series of seismographs also placed on the bottom. These seismographs and their records were later recovered by freeing weights, which held the instrument down until a salt release dissolved in the water, dropped the weights and allowed the buoyed instrument to rise to the surface. This was a costly method and has been largely replaced since it was learned that explosives set off near the surface are almost as effective.[2] With this innovation floating seismographs were strung out behind the ship, and the records were received in each of them, but it was soon found more satisfactory to set off a series of explosions from one moving vessel, with a stationary vessel receiving the sound waves on its seismographs. The distance between the vessels at points where explosives are dropped is obtained by underwater sound.

The refraction method not only has given much information on the thickness of unconsolidated sediment but also has indicated the presence of several layers underneath the sediments in which sound velocity changes abruptly. The most notable achievement of this refraction method has been the determination of a large discontinuity underlying both the ocean and the continents.

Thickness of Deep-Ocean Sediment Cover

Many geologists have speculated on the thickness of the sediment that blankets the rock of the ocean bottom. The most painstaking attempt to arrive at a satisfactory average through use of the various methods was made by Philip Kuenen in his *Marine Geology*. Using information from various cores that showed the contact between glacial age deposits and post-glacial, he arrived at an average rate of accumulation of one centimeter (0.4 inch) per 1,000 years. He reasoned that a slower rate should apply to past ages because of the supposed greater elevation of the continents during the Recent period than during most of the geological ages, and because of compaction of deeply buried sediment. Applying corrections, he arrived at a figure of 1/6 centimeter per 1,000 years. Then figuring on two billion years for the age of the ocean, he estimated that the average thickness for the sediment of the ocean basins would be three kilometers (9,800 feet) or approximately two miles. Using various other methods, such as multiplying the ocean age by the estimated amount of sediment carried annually by rivers from the continents to the ocean, Kuenen obtained figures that were roughly comparable. As a result, it was anticipated that when the sediment thickness could be measured by sound waves, the results would confirm Kuenen's estimates.

We now have many sound records, but little that will confirm Kuenen. We know that the approximate thickness[3] in the low-velocity layer down to the first good reflecting surface in many parts of the Pacific averages about 1,000 feet, and in the Atlantic it averages about 2,000 feet. The higher values in the

---

[2] The sea-bottom method has been revived recently by John I. Ewing of Lamont Geological Observatory for special purposes.

[3] Approximate because the velocity of sound in unconsolidated sediment varies with porosity and compaction so that an assumed value has to be given. This, however, is probably of the right order.

Atlantic were expected because the rivers carry more sediment into that ocean per unit area of ocean than into the Pacific. In any case there is a great discrepancy between the Kuenen estimates and the measured values.

As yet no one has entirely explained the difference between the estimates and measurements. It is possible that the measurements have been made in areas where sediments are much thinner than the average, but this does not seem reasonable because the relatively thin sediment cover has been found in both the Atlantic and the Pacific. An alternative could be that the oceans are much younger than is generally thought to be the case. The great submergences indicated by the deep guyots, by the deep foundation of atolls, and perhaps also by the great depths of submarine canyons may imply that the ocean has been growing deeper rather rapidly since the Cretaceous period, as suggested by Roger Revelle. A rise of 5,000 feet of the sea surface during the past 100 million years[4] might mean that the ocean is less than 250 million years old (the average ocean depth being 13,000 feet). This would account for a thinner sediment cover.

An explanation that appears to offer fewer difficulties has been suggested by E. L. Hamilton. His suggestion is that the deeper parts of the sediment cover have been converted into rock, so that the reflection or refraction that comes at the base of the supposed sediment column is actually at the contact between the unconsolidated sediment and a consolidated sedimentary layer. If there is a sequence that consists of a consolidated layer lying between unconsolidated sediments above and below, the existence of the underlying sediments cannot be detected by sound impulses, since the refraction of sound will be upward only where the sound passes from an unconsolidated into a consolidated layer (Fig. 90). Perhaps a thin layer of lava has poured out over the sediments in some places and hidden the underlying portion. If this explanation by Hamilton is correct, we would have less difficulty in interpreting past geological history than with the hypothesis of a relatively young ocean. According to most geologists, the ancient formations and the fossil record seem to require an ocean throughout geological history.

Thickness of Marginal Sediments

In some places along the margins of the ocean basins there are thick masses of sediment. Off the coast of the United States the Lamont group has found two sedimentary troughs, an inner one under the continental shelf containing an estimated thickness of 17,000 feet of sediment and an outer one under the slope and adjacent ocean floor with up to 30,000 feet of sediment (Fig. 91). Similarly, along the United States coast of the Gulf of Mexico there are 20,000

[4] The approximate age of the Cretaceous.

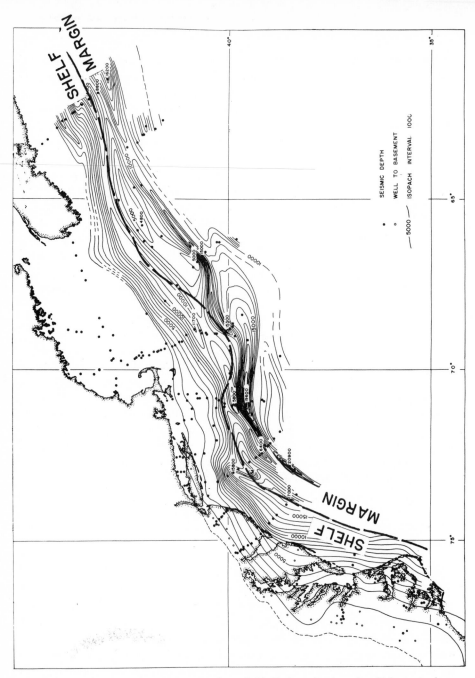

Figure 91. Contours from C. L. Drake and M. Ewing showing the thickness in feet (isopachs) of the unconsolidated and semiconsolidated sediment off the east coast. A rise in the basement rock near the edge of the continental shelf is indicated. Courtesy of Lamont Geological Observatory.

feet or more of sediment under the outer shelf according to the reports of various petroleum geologists. Off California, Scripps Institution scientists have found as much as 10,000 feet of sediment under the basins of the continental borderland (Fig. 92), but elsewhere, particularly on the ridges, rock comes right to the surface. Beyond the continental slope the sediment is more like that of the deep-ocean basins, although locally it may thicken up to about 6,000 feet.

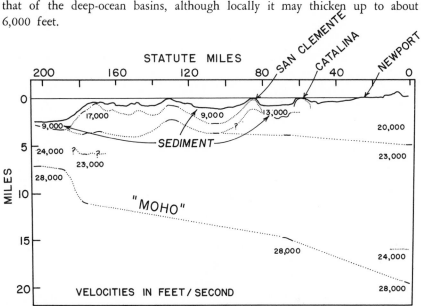

Figure 92. Cross section by George Shor and R. W. Raitt, Scripps Institution of Oceanography, of the sea floor off Newport, southern California, showing the thickness of the sediment under the basins and the travel velocities in feet per second in the underlying rocks. The Moho disconformity is shown at the base. Note that the Moho is far deeper under the continent than under the ocean basin.

Crustal Layers Under the Ocean

The crust and mantle. The interpretation of refraction shooting over the ocean basins soon led to the recognition of a major discontinuity at a depth of from about three to eight miles below the bottom, averaging four miles. Below this break, sound travels at a rate of a little more than 26,000 feet a second and above the break at various slower speeds. The contact is called the *Mohorovičić discontinuity,* or the *Moho* for short. It occurs also under the continents, but here it lies at a depth of about twenty to twenty-five miles, a striking contrast with the oceanic values. Above the Moho the rock is generally referred to as the *crust* and below as the *mantle.*[5]

[5] There are other deeper layers, which will not be considered here because they are the same under ocean and continents.

The discovery of the shallow Moho under the oceans has been very helpful in the explanation of the ocean depths. The rock below the Moho with its higher sound velocity is best interpreted as heavier than the rock above, since denser rocks in general have higher sound velocities. Therefore, the thicker crust under the continents with its lighter weight can balance the thinner crust under the ocean.

The Moho discontinuity is difficult to find in some places under the Mid-Atlantic Ridge. At these localities there appears to be a gradational zone between the crust and the mantle. Sound velocities a little low for the mantle but high for the crust extend to depths that are in excess of the usual depths of the Moho under the ocean. It has been suggested that under the broad arches of the ocean floor the crust and mantle may have been mixed together by the development of molten rock along the contact.

At first it was thought that the crust under the deep oceanic trenches would be still thinner than under the rest of the ocean in order to balance these great depressions. Recently some measurements have been made, mostly by Scripps Institution geophysicists, that show the depth of the Moho under trenches of the Pacific. The Moho has been found to bend down under the trenches rather than up, which may imply that the trenches are being held down by some lateral force, possibly because they are so narrow and therefore can support the weight.

The contact between the thin oceanic crust and the thick continental crust is still not entirely clear. However, there is some information that indicates that the continental shelf lies in a transition zone with the Moho dipping rather steeply toward the continent under the ocean (Fig. 93). The continental borderland off southern California, which has somewhat intermediate water depths between oceanic and continental, also has intermediate thickness of the crust above the Moho (Fig. 92). George Shor and Russell Raitt found that the Moho lies closer to the surface along the outer borderland than it does

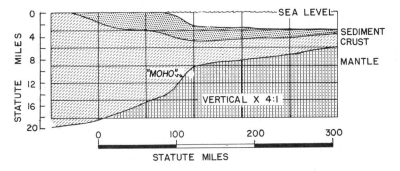

Figure 93. Generalized diagram showing the relation of the overlying sediment, the crust, and the mantle for the east coast by J. L. Worzel and G. L. Shurbet, Lamont Geological Observatory. Note that the same relations exist as are found off the coast of California (Fig. 92).

under areas of the same depth on the inner portions of the borderland. Thus the crust thickness here, as under the trenches, is not directly related to water depth.

Another interesting result from the refraction studies has come from the investigation of relatively small basins like the Gulf of Mexico and the Arctic Ocean. It has been thought by many geologists that the Gulf of Mexico is a sunken portion of the North American continent, but the Moho in all of these relatively small basins of considerable depth is almost as near the bottom as under the large oceanic basins. It is difficult to believe that a continental mass with its deep Moho could be transformed into an ocean basin with its shallow Moho. Such a change would have to include a transformation of some 20 miles of crust into rock like that of the mantle. It seems more likely that the Gulf of Mexico has always been deep and hence that the geological interpretations are in error, as they could easily be.

Intermediate crustal layers. In the early interpretations of refraction shooting over the ocean floor only one crustal layer was recognized under the sediment cover. Later, however, the extensive work by Russell Raitt and his colleagues in the Pacific showed that there was definitely another layer. This same layer has since been discovered in the Atlantic by various investigators. As it now stands there is a downward sequence as follows: the sediment layer with an average velocity of 7,000 feet per second, the first rock layer with an average velocity of 16,000 feet per second, and the deep crustal layer with an average velocity of 22,000 feet per second. The upper rock layer could be either sedimentary rock, such as limestone, or lava. Around volcanic islands, some of the velocities are as low as 16,000 feet per second, and these are almost certainly lavas with many gas cavities, or the rocks are highly fractured. The velocities of 21,000 to 23,000 feet per second found in the deep layer indicate that it is an igneous rock, more likely a crystalline rock, such as gabbro, than the fine-grained basalt found in many volcanic islands.

Gravity Interpretations for the Ocean Basins

Another method of diagnosing the nature of the rock under the oceans is by measuring the pull of gravity. This has been accomplished by timing the swing of a pendulum in a submerged submarine, where there is a minimum of wave disturbance. At any given spot the swing of a pendulum has a constant period. The greater the pull of gravity, the faster the pendulum will swing. It is necessary, however, to get a very accurate timing of the period, since differences are only a few parts in a million. In addition, the topography must be taken into account. The higher the elevation at which the pendulum is swung, the farther the station lies from the center of mass and hence the slower the swing. Thus corrections have to be made for topography even though all ocean measurements are made at almost exactly the same height. However, in oceanic gravity measurements a correction has to be made for depth of water, since water is lighter than rock and will therefore not have as much gravitative pull. Once these and some other corrections have been made, the reading of gravity gives an idea of the relative heaviness of material underlying the area where the measurement has been made.

Gravity studies in the ocean have shown that in all probability the oceanic crust as a whole is balanced against the continents just as a heavy piece of wood floating in a pail of water is balanced against a light piece of wood, the former rising less above the water surface. The continents rise above the ocean-bottom level because they are underlain by lighter material. The gravitative studies are essentially in agreement with those of explosive waves and earthquake vibrations, since in general the heavier rocks have also higher speeds of wave travel. The gravity measurements are used to help determine the thickness of crustal layers, work in which Vening Meinesz of the Netherlands and in recent years J. Lamar Worzel of Lamont Geological Observatory have been leaders.

One of the interesting results of gravity measurements has been the finding of areas with decidedly deficient gravity wherever the deep-sea trenches have been investigated. This has led to the speculation that the trenches may be held down by lateral pressure, due perhaps to convection currents that move the crust and produce mountain ranges in other places. A difference of opinion has developed regarding the interpretation of the gravity anomalies and the crust under the trenches. The Lamont group maintained on the basis of gravity measurements that the crust is very thin under the trenches, but as indicated previously the seismic shooting by Scripps Institution scientists indicates that the crust is relatively thick and the Moho bends down under the trenches.

Heat Rising from the Ocean Floor

Geologists have long been interested in the heat flow from the interior of the earth. At first it was supposed to represent the heat from a cooling molten core. Before the heat-producing effects of radioactivity were appreciated, it was believed that the earth could not be more than about 80 million years old or the interior would have lost all of its heat. Then with the discovery of radioactive substances in crustal rocks and the study of their rate of heat generation it was learned that enough heat is being produced to keep the interior at a high temperature for billions of years. This discovery was very helpful in the interpretation of volcanic activity through the geological ages. According to the old idea, most of the volcanic activity would be ancient, but we know that tremendous volcanic outflows have occurred in relatively recent geological periods producing, for example, the great lava plateaus of Washington and Oregon and the even greater flows of central India and Tibet.

Study of radioactivity has developed some puzzling features. For instance, it was learned that the granitic rocks of the continents have many more radio-active minerals than the basalts of the volcanic islands. This led to the thought that radioactivity was more prevalent under the continents than under the oceans. In turn this would suggest that there would be more volcanic activity on the continents than in the oceans. However, there appear to be many more volcanoes rising above the ocean floor than above the continents. This is partly explained by the erosion of continental volcanoes and the preservation of those in the oceans. But even allowing for this differential it seems likely that vulcanism is at least as active in the oceans as in the continents.

Measurement of heat flow from the continents is difficult because of the fluctuation of temperatures at the surface. As a result, temperatures have to be measured in deep wells below this fluctuation and below the effect of circulating waters. Here there is more difficulty in preventing the penetration of surficial temperatures because of circulation of air and water in the well. On the deep-sea floor, on the other hand, the temperature is constant, so that no trouble arises from these sources. Long probes containing recording temperature elements at both top and bottom have been used to penetrate about ten feet into the ocean bottom and make a record of the temperature while in place. These can be read when the instrument has been brought back up to the surface. Cores are taken from the adjacent sea floor, and the heat conductivity of the material is measured and used to apply corrections to the readings.

The measurement of heat flow from the ocean floor in both the Atlantic and the Pacific has now been made in many localities. The chief credit for this work goes to Sir Edward Bullard, Arthur Maxwell of the Office of Naval Research, and Roger Revelle. They have found that instead of heat flow being decidedly less from the ocean floor than from the continents, it is if anything slightly greater. Furthermore, it appears to be more variable under the ocean than under the continents, although measurements under the continents are too scarce to make this very sure, particularly since they come from oil wells and mines that are not very typical. Perhaps the most interesting finding for the oceans is that the heat flow is very high in the East Pacific Rise of the eastern equatorial Pacific and in parts of the Mid-Atlantic Ridge. On the other hand, it is quite low in some of the trenches of the eastern Pacific. It seems possible that the reason the elevations of the rises are greater than those of the adjacent ocean floor can be explained in part by the greater temperature of the underlying crust. Several scientists have suggested that the higher temperature is the result of convection currents (like those in the water of a teakettle), with a slow rise of the crust taking place under the great oceanic ridges and outward movement on either side. This brings hotter rock up into the top of the ridge because of the outward flow, exposing more deeply buried layers.

This hypothesis of convection currents is also used by many scientists to explain the major movements of the earth's crust and even the development of mountain ranges. It is perhaps wiser to say that the present state of knowledge of the oceans' foundation is sufficient only to eliminate some old hypotheses. It is scarcely adequate to establish new hypotheses with any assurance that they will remain permanently.

CHAPTER X

CORAL REEFS AND THEIR UNDERSEA WONDERLANDS

Until Jacques Cousteau and other talented underwater photographers had revealed their remarkable pictures, few people realized that coral reefs contained what could easily be described as the most magnificent and awe-inspiring scenery on the earth. The tremendous branching corals with their grottoes and narrow passageways, where myriads of brilliantly colored fish form a never-ending parade of beauty, are simply beyond description. Pictures, no matter how good, fail to provide an adequate substitute for seeing in person these marvels of nature. My first opportunity to examine a reef came during World War II when I spent several weeks on a coral island. I was told by various members of the local military personnel that it was just like being in jail to be stationed there. Instead, having a face mask and a week's delay in receiving necessary scientific equipment to get to work, I had one of the most glorious experiences of my life exploring the local underwater scenery.

The submerged coral reefs and the low coral islands associated with them owe their existence to a strong but flexible framework of branching colonies of calcareous animals and plants, of which the corals are the most important and most impressive supporting members. Inside the frame, the skeletons of small animals and plants have been deposited and form the bulk of the reef mass. The whole structure is reminiscent of a filled corn crib with its frame supporting the ears of corn that are piled up in the interior.

The coral reefs are largely a monopoly of the tropical portions of the Pacific Ocean, where there are literally hundreds of them scattered over an area extending for 6,000 miles in a northwest-southeast belt that has a width of about 1,500 miles (Fig. 94). They occur elsewhere but more sparingly. Notable among the other reefs are the extensive Seychelles and Maldive Islands reefs of the Indian Ocean, the colorful reefs of the Red Sea, and the relatively small reefs of the West Indies and Bermuda.

History of Exploration

Long before the coming of the white men the coral islands were inhabited by Polynesians, Melanesians, and other venturesome peoples coming from Asia in their not too primitive canoes. The reefs surrounding the volcanic islands provided a protection for landing the native boats. The lagoons inside the reefs also contained calm waters in which the natives could fish and obtain almost enough food for their sustenance. In contrast to the volcanic islands surrounded

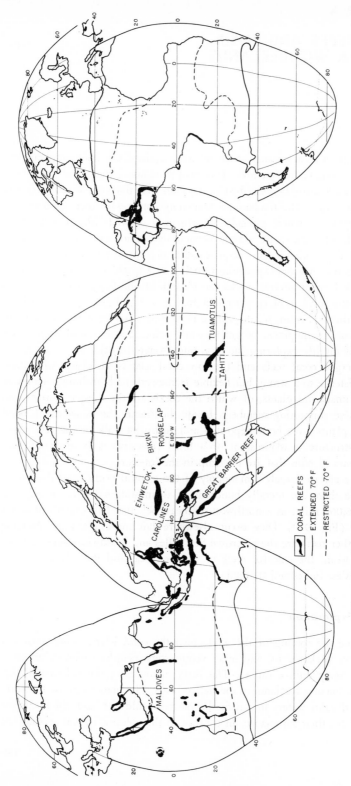

Figure 94. The distribution of coral reefs. Areas of the reefs are exaggerated because of scale. Isotherms show the belt that has 70° temperatures in summer "extended" and in winter "restricted." Adapted from John Wells, Cornell University.

by reefs, those islands built entirely of coral proved less secure for the natives because of their low elevation, which permitted large waves to come over them during great storms.[1] Nevertheless these low islands were also widely inhabited. The first crossing of the Pacific by white men, accomplished by Magellan in 1520, unfortunately missed all the coral islands, but they were soon discovered by his successors and were visited many times during the first half of the sixteenth century. Captain Cook, in his fruitful trips at the time of the American Revolution, was among the first to bring back important facts concerning the nature of the Pacific coral reefs, but it was not until Charles Darwin made his voyage on the British vessel *Beagle* in 1835 that a documented study of some of the reefs was forthcoming. Shortly after the long voyage, Darwin published his ingenious theory of the subsiding volcanic mountains and upgrowing reefs, which was debated for more than a century thereafter.

Much of the exploration of the reefs following that of Darwin seems to have lacked his keen perception and his energetic search for facts. In the first half of the twentieth century two outstanding American scientists, William Morris Davis and Reginald Daly, both of Harvard University were writing and lecturing about coral reefs and carrying on a heated debate about whether the Darwin subsidence hypothesis or the idea of glacial swinging sea levels, first suggested by Albrecht Penck, was the correct explanation. Davis favored Darwin, and Daly the Penck hypothesis, which he developed so extensively that finally it became generally accredited to him. Despite all of their debating, neither of these world-famous scientists studied the reefs except from the literature and from brief tours through the coral islands, comparable to the typical American tour of Europe.

Meantime a number of scientists were slowly gathering facts about the reefs, although in a less spectacular fashion.[2] The real turn in the tide came with a surge of progress after World War II when the United States Navy's Joint Task Force I sponsored a series of studies in the Marshall Islands in connection with atomic and hydrogen bomb tests. These carefully planned expeditions represent the first large-scale investigation of the nature of coral reefs by a well-organized team.[3]

[1] *The Hurricane* (Boston: Little, Brown & Co., 1936) by Charles B. Nordhoff and James Norman Hall provides a vivid and realistic description of such a storm.

[2] Among these were T. Wayland Vaughan, Harry Ladd, and J. Edward Hoffmeister of the U.S. Geological Survey, A. G. Mayor of the Carnegie Institution of Washington, C. M. Yonge of Glasgow University, and J. Stanley Gardiner of Cambridge University.

[3] Credit for extensive results of this work goes to Kenneth O. Emery of Woods Hole Oceanographic Institution, Roger Revelle of Harvard University, and Harry Ladd and J. I. Tracey, of the U.S. Geological Survey. Other important studies have been made by Preston Cloud of the University of California, Los Angeles, John Wells of Cornell University, and Norman Newell of Columbia University and the American Museum of Natural History. Newell has been greatly assisted by major oil companies whose geologists became interested in modern reefs because of the discovery that ancient reefs contain a large, perhaps the largest, source of petroleum for the future.

The Nature of Coral Reefs

Coral reefs, despite their diverse forms, have certain characteristics in common. All of them are rocky mounds, platforms, or ridges rising slightly above the general level of the adjacent sea floor and consisting largely of the skeletal remains of organisms. The outer frame, which holds the mass together, consists mostly of branching corals (Fig. 95) that have grown up from the sur-

Figure 95. A Lithothamnion ridge rising above the outer margin of the reef. Photograph courtesy of U.S. Geological Survey.

rounding surfaces by new generations growing on top of the skeletons of the old. Inside the frame, various types of debris have been introduced by waves from the outer reef. This accumulates along with growing corals (Figs. 96, 97), shells of organisms, and the creeping, lime-secreting plant known as Halimeda (Fig. 99) and the encrusting algae called Lithothamnion (Fig. 95). The latter also grow at ridges around the margins of reefs.

A coral reef is by no means a solid mass of rock. The coral often grows across a depression in the reef surface, leaving a cavern beneath. Some of the holes

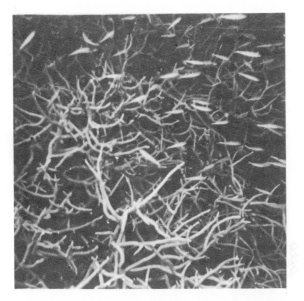

Figure 96. A thicket of staghorn coral, a variety of Acropora, in Bikini Lagoon. Photograph courtesy of U.S. Geological Survey.

Figure 97. The umbrella-like coral, a variety of Acropora, growing in ninety feet of water in the South Pacific. Photograph by John MacFall, Scripps Institution of Oceanography.

in the reef are filled with debris, but some are simply pockets filled with sea water. Smaller holes in the reef are the result of boring organisms, such as the spiny echinoids (Fig. 98) and the boring molluscs. These organisms are destroying reefs, but their work is more than offset in places where the corals and algae are prospering.

Figure 98. In lower right of photograph are spiny echinoids, which burrow into coral masses. The coral next to the scuba diver is called Heliopora, and the coral at the lower left is Porites. Photograph by Willard Bascom.

Conditions Necessary for Coral Reefs

Whereas many solitary corals can live in almost any ocean temperature and over a very wide range of depth, the reef-forming corals are much more restricted.[4] The reefs are found only in areas that have warm water during most of the year. It is often stated that coral reefs cannot survive winter temperatures below 65° Fahrenheit or, at the least, 61°. This may be essentially the case, but some recent records of temperatures made by the Aramco operations near Dhahran in the Persian Gulf show that at least during the past five years some small reefs have survived temperatures as low as about 52°. Possibly this is balanced by the temperatures in the low 90's that occur in summer. On the other hand, temperatures over 96° are probably lethal to most corals. The limiting depth of active reef growth is usually about 150 feet, although some live reef corals have been dredged from as deep as 580 feet.

The corals are also dependent on relatively saline water, i.e., from twenty-seven to forty parts per thousand (normal sea water has thirty-four to thirty-six parts per thousand). They are sometimes killed by unusually heavy rains, which greatly reduce the salt content. At Pago Pago in American Samoa, according to A. G. Mayor of the Carnegie Institution of Washington, a great downpour occurred in 1920 with a total of thirty-seven inches of rain falling in four days. This killed all of the local reef either by diluting the sea water or by washing mud onto the reef. Corals grow much better in clear water than in turbid water, but they can stand protracted periods of water so muddy that one can see bottom only in a foot or even less. I know this from experience because, while swimming under normal weather conditions in south Molokai, I bumped my knees on a live coral that I could not see through my face plate until I was within a few inches of it.

The food supply is of course very important in coral-reef growth. The carnivorous corals capture the minute animals (zooplankton) drifting past them by stinging them and hauling them in with their tentacles, which they extrude at night. The corals are therefore dependent on water circulation. This is evident from the better growth of corals observed on the windward side of islands and continents than on the leeward. The winds drive currents more toward the windward shores and hence provide food for the corals. However, here again there are exceptions, and some of the leeward shores of islands have better reefs than those of the windward side on the same island. This is true of the islands of Hawaii and Molokai in the Hawaiian group, but Oahu and Kauai have their best reefs on the windward side. The difference is apparently due to some peculiarities of the local currents.

---

[4] Most of the corals in coral reefs are *hermatypic,* that is, they have symbiotic flagellates. There are ahermatypic corals that may form small reefs in deep cold water, like those described for the outer edge of Blake Plateau (p. 100). The limiting conditions referred to in this section are for the typical reef-forming corals.

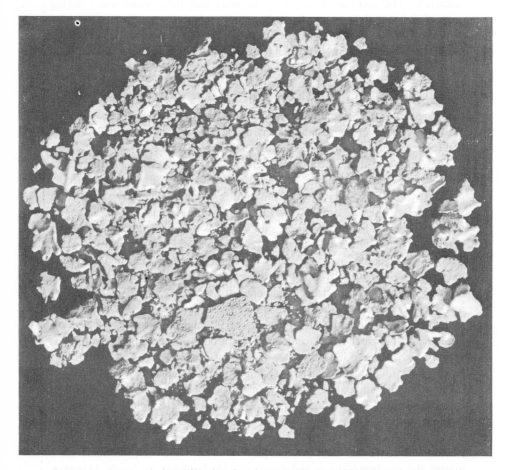

Figure 99. A mass of the Halimeda taken from the floor of Bikini Lagoon. Photograph courtesy of U.S. Geological Survey.

Types of Reefs

The three dominant types of coral reefs are *fringing, barrier,* and *atoll* (Figs. 100, 101, 102). Fringing reefs border the coasts of many islands. In the Hawaiian Islands you can wade out on reefed platforms for distances of as much as a mile. Most fringing reefs have few growing corals and are

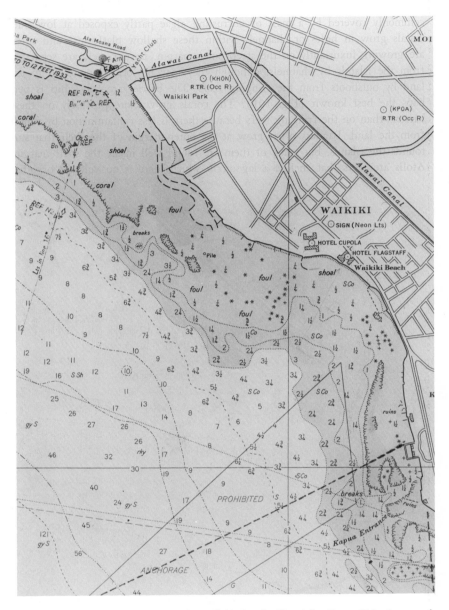

Figure 100. Fringing reef west of Waikiki Beach, Honolulu. From U.S. Coast and Geodetic Survey Chart 4132.

commonly covered by a mass of encrusting algae partly exposed at low tide. Corals grow in holes or channels within these shallow platforms[5] and often are growing luxuriantly on the steep outer margin. Barrier reefs are separated from the islands by channels, although they may be locally connected to the land by outshoots from the fringing reefs. The barrier around Tahiti is perhaps the best known (Fig. 101). The corals grow more profusely on these barriers than on the fringing reefs because there is less contamination by runoff from the land. Barriers often grow at the outer margin of the shelf that surrounds the islands, but many of them are found well inside the shelf margin. Atolls are reefs that enclose a lagoon in which there is no volcanic or other

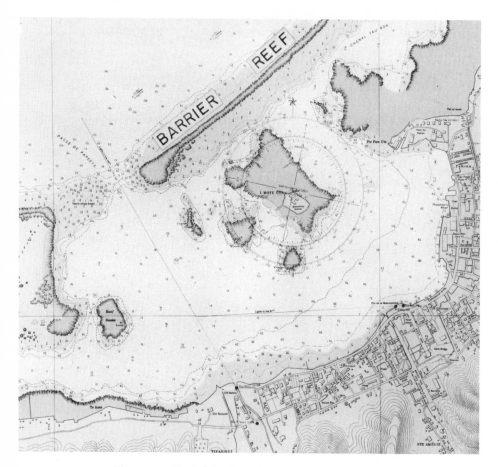

Figure 101. Typical barrier reef, Papeete Harbor, Tahiti.

[5] A few hundred yards off Waikiki Beach, just west of the surfing area, you can find some corals still growing in these holes despite souvenir hunters.

high island. Atolls are often thought of as ring shaped, like Eniwetok in the Marshalls (Fig. 102), but actually the great majority are quite irregular in shape, most of them having prominent points and indentations.

There are other types of reefs that are also important. Among these are *table reefs,* which are flat-topped isolated masses that constitute dangers to navigation in many parts of the tropical oceans. Some of the former table reefs

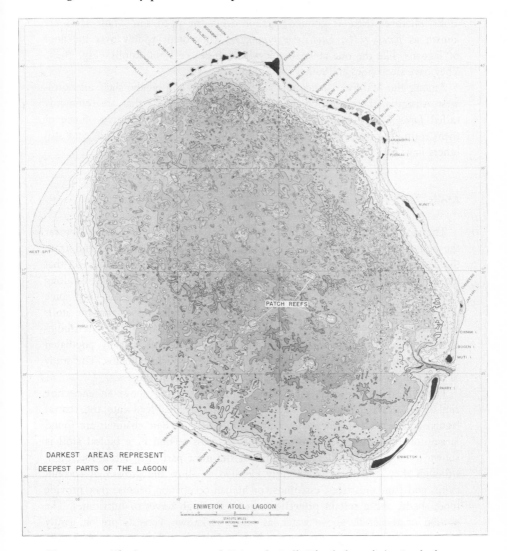

Figure 102. The bottom contours of Eniwetok Atoll. The darkest shades in the lagoon indicate depths of 192 to 216 feet. The channel entrance on the windward side of the island is exceptional. From K. O. Emery, J. I. Tracey, Jr., and H. S. Ladd, U.S. Geological Survey Prof. Paper 260A, 1954.

have been elevated and constitute flat-topped islands, like Washington Island just north of the Equator near Hawaii. Others are deep enough that they are not a danger to navigation, like Munyal Bank west of Mangalore, India, with its 120-foot minimum depth. Most of the table reefs, however, are so shallow that they are difficult to investigate, because they have waves breaking over them much of the time.

Another type is the small oval-shaped reef that grows quite profusely inside atolls wherever there is good circulation from the open sea. These reefs are known as *patch reefs, knolls,* or *pinnacles.* They occur widely over the floor of lagoons, like the one at Eniwetok where there are 1,500 knolls (Fig. 102). They have steep sides and rather small tops.

Among the Maldive Islands in the Indian Ocean and on the shelf off north-west Australia, there are numerous small ring-shaped reefs that are commonly called *faros* (Fig. 103), a local word used by the Maldive natives. Some of them form crescents, others have complete rings with a lagoon inside, and still others have had their lagoons filled but have a slightly raised rim.

Atolls

The atolls are the most common type of coral reef in the oceans and by far the best known. The hundreds of atolls in the equatorial Pacific extend from the Tuamotus on the east to the Carolines on the west (Fig. 94). The fact that most of these islands have no sign of volcanic rock either in the center or along the rim has aroused much interest among scientists because corals need some base on which to grow. The desire to find this base is one reason that the atolls have been studied so much. The other more urgent reason for recent studies is that the atolls are located in a sufficiently isolated and sparsely populated area to favor their selection as a testing ground for nuclear bombs. The study of the reefs was necessary to establish the full effects of the tests.

The typical atoll consists of a chain of islands rising above an encircling rim of shoal water with a few transverse channels that lead into the central lagoon and provide an access for vessels (Fig. 102). These channels are found largely on the leeward sides of the atolls. A cross section of a typical atoll is shown in Figure 104. The islands are for the most part only a few feet high, although they usually are sufficiently elevated to allow the growth of coconut palms and pandanus trees, essentials for human occupation because they provide food, shade, and a certain protection against the high waves of hurricanes. The portion of the atolls above water, as far as is known, consists predominantly of masses of rubble from the surrounding reef that have been thrown up by the waves into a rampart. As would be expected, these islands are most developed on the windward side of the reefs. In addition some elevated platforms represent a former reef that grew below sea level and has been subsequently elevated.

Figure 103. Faros, small atoll-like reefs, superimposed on a large atoll in the Maldive Islands. Depth in fathoms.

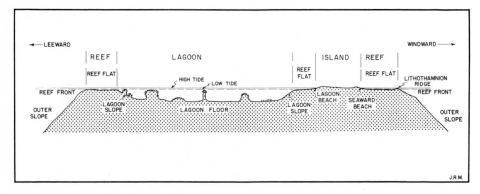

Figure 104. Diagrammatic cross section of an atoll. After J. I. Tracey, Jr., of the U.S. Geological Survey, P. E. Cloud of the University of California, Los Angeles, and K. O. Emery of Woods Hole Oceanographic Institution. Vertical scale exaggerated.

Most of these do not stand more than about three to five feet above sea level, and they are relatively rare, at least among the atolls of the Pacific. Some of the islands have dunes constituting their highest elevations. The dune sand has no quartz or any other minerals that are found in typical dunes on the mainland. The coral island dunes consist of fragments of coral, molluscs, algae, and bryozoa, along with many foraminifera.

On the outside edge of the typical island there is a beach of coral and shell sand or gravel bordered by the shallow reef flat. The latter consists largely of dead coral or algae, but has patches of living coral occurring as low mounds or lining the sides of channels that cut across the flat. In many places this reef flat shows a gradation outward from the massive Porites coral along the shore, to the blue Heliopora, and then to the more delicate branching Acropora outside (Fig. 96). Beyond the reef flat there is often a ridge consisting of the reddish algae known as *Lithothamnion*. This plant, unlike the coral, can grow above low-water level, being supported by splash and spray.

Beyond the ridge is a slope in which there are many grooves, especially on the windward side of the islands. It is easy to spot these from the air (Fig. 105). Some of these grooves extend deep into the reef flat, being partly roofed over by the growing algae and coral. These inner passageways are a fascinating place for exploration for scuba divers. They are inhabited by beautiful colored fish of all shapes and sizes, and the walls of some of the passages have a luxuriant growth of corals.

At the base of the first slope small terraces are found, often about ten fathoms deep. Beyond that the slope is quite continuous down to deep water, commonly having an angle of inclination of close to 25°. This is thought to be the steepest angle that is stable for coral debris. On the leeward side of some

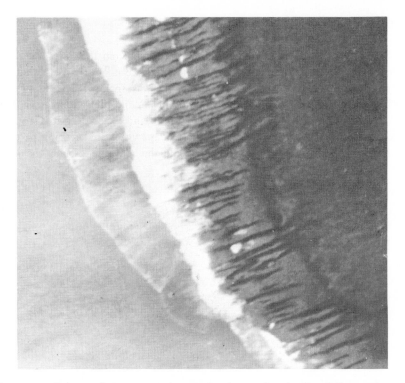

Figure 105. Submerged grooves on the margin of a coral reef. The shallow reef is on the left and deep water on the right. Photo by R. Dana Russell.

of the atolls, however, the upper oceanic slopes are often vertical or even over-hanging. These cliffs are honeycombed with caves, and large blocks break off from them and lodge on the lower slopes. Farther down the atoll slopes, the bottom changes gradually from a mass of rubble blocks to a coral sand zone with few blocks and finally into a fine coral debris.

On the lagoonward side of the atoll there is another shallow platform, which is likely to be narrower than that on the outside. A gentle slope leads to the floor of the lagoon. The floor, contrary to some of the literature on reefs, is by no means as "flat as a billiard table." Almost all atolls have numerous patch reefs and faros rising above the floor. Furthermore, the floor is not perfectly level but has numerous shallow basins and small rises (Fig. 102). In some places, however, fathograms show broad stretches with little relief. Here, samples of the bottom show a combination of fine reef debris, with the vinelike plant Halimeda (Fig. 99) and many foraminifera. The passes commonly leading into the lagoon have depths that are comparable to those of the deep parts of the lagoon.

Origin of Atolls

Since we know that coral reefs cannot grow up from the deep-ocean floor, the atolls as well as the table reefs of the open ocean must have started on a pre-existent foundation over which the water was shallow. The simplest explanation is that proposed by Darwin (Fig. 106), in which the corals started as fringing reefs around a subsiding volcanic island. According to Darwin, as the island subsided, the outer part of the reef grew upward, keeping pace with the sinking, so that a lagoon formed inside the reef and around the island.

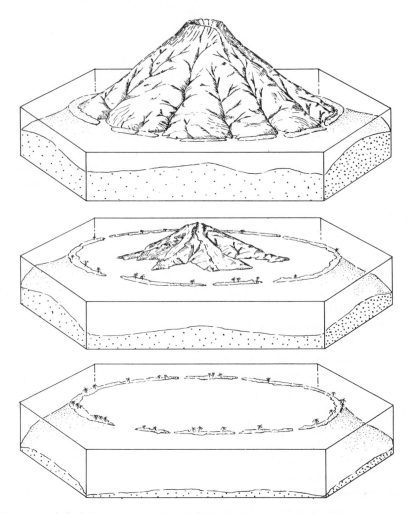

Figure 106. Block diagrams to indicate Charles Darwin's hypothesis of the development of an atoll by the submergence of a volcanic island and the upgrowth of a reef.

Eventually, the entire island disappeared, and the continuously upgrowing reef then formed a ring with only a lagoon on the inside.

This explanation failed to account for some facts, most of which were learned after Darwin's memorable voyage. One of these is that lagoon floors are all quite shallow, whereas very deep moats should have developed as a result of the sinking of large volcanic islands. Extensive atolls, such as Eniwetok, Rongelap, Bikini, and Kwajalein (all some twenty miles across), ought to have a deep moat inside the reef, shoaling toward the center of the lagoon. Partly because this had not been found, Reginald Daly adopted the idea of Albrecht Penck that the atolls were the result of upgrowth of reefs during the postglacial rise of sea level. Daly developed this changing-sea-level hypothesis much further and suggested that the cold glacial stages were accompanied not only by lowering of sea level but also by a chilling of the waters in the coral-reef areas, which killed many of the reefs. He thought also that the waters became muddy because of the stirring up of coral muds exposed by the lowered sea levels and that this helped kill the reef corals. With the extinction of the reefs the waves were free to attack the unprotected islands, so that wave-cut platforms were formed (Fig. 107). When the sea became warmer and the sea level rose because of the melting of the ice caps, the reefs grew up on the edge of these platforms.

The Darwin proponents tried to explain the absence of a moat in the lagoon floor as the result of fill from the outside as well as of sedimentation and coral

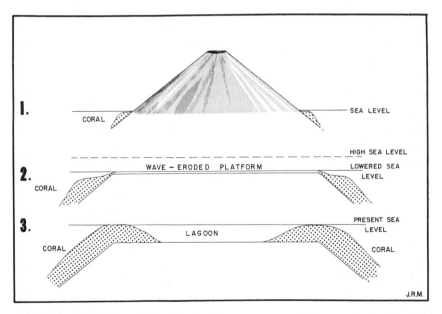

Figure 107. Daly's hypothesis of the development of an atoll by wave erosion during a low stand of sea level and upgrowth around the rim when the sea level rose.

growth on the inside. They countered Daly's arguments of glacial stage platforms by referring to the immense size of many of the lagoons. Obviously it
would take a long time to bevel the large volcanic islands and produce the
necessary platforms. The glacial episodes were too brief for this.

Drillings in the atolls. By drilling into the atolls and studying the drill cores
it was possible to test the relative merits of the two hypotheses. The first of
these drillings was made in 1897 when the British put a hole down into Funa-
futi Atoll to a depth of 1,100 feet. If they had encountered volcanic rock at
shallow depth, that would have been substantial evidence for the Penck-Daly
hypothesis. Actually they never got through coral material, so that the Darwin
school considered it a victory for their side. Daly, however, argued that the drill
had merely penetrated coral talus built out beyond the reef, as in Figure 107.
A drilling in Bermuda in which the coral cap was found to be only 245 feet
thick was considered to be favorable to glacial control.

Other drillings have followed, including one to 1,400 feet by the Japanese
just before World War II in Borodino Island 200 miles east of Okinawa, and
one to 1,600 feet by the Dutch on Maratua Island off Borneo. Both failed to
get to the bottom of the coral. Drillings into Bikini in 1947 to a depth of
2,500 feet again failed to get through the coral. However, Daly still claimed
that it was only talus that had been penetrated.

A really large-scale operation was needed. This was undertaken in 1952
at Eniwetok by the United States Navy and United States Geological Survey
under the direction of Harry Ladd of the Geological Survey. Holes on both
sides of the atoll went down to 4,600 and 4,200 feet, where they both encountered lava. Even here the claim could have been made that the section
was simply talus. However, these core samples have been studied with great
care, as have also those from the Bikini boring. They show conclusively that
shallow-water conditions existed throughout the time when these thousands
of feet of coral formations were being deposited. It would be easy to recognize
talus deposited on the deeper portions of the atoll slope because the interstitial
material would contain many deep-water foraminifera. Only a few intermediate-depth forms have been found. Furthermore, at some points along the drillings, cores were taken of large coral masses that appeared to have been growing
in place, although this is difficult to ascertain. The frequent encountering of
cavities, one as much as 55 feet across, showed that the drilling must have been
through a reef rather than a deep talus slope, where no holes of any appreciable size could have existed. Also the drilling shows an orderly sequence of
deposits from the Eocene, some 60 million years ago, up to the present day.

The other drillings show the same sequence, although their significance has
not been entirely appreciated. Even the Bermuda drilling does not favor the
glacial-control hypothesis because the top of the volcanic rock instead of having
the fresh appearance of a wave-cut platform is deeply weathered down as

far as 455 feet, which must have been a result of a long period during which the island stood much higher than the present.

All of this new evidence is inclined toward Darwin's subsidence hypothesis, but it does not indicate the upgrowth of a reef around a subsiding volcanic cone. A more likely interpretation from the facts now at our disposal is shown in Figure 108. We can assume that at some time in the remote past, possibly during the Cretaceous period, some 100 million years ago, there were many

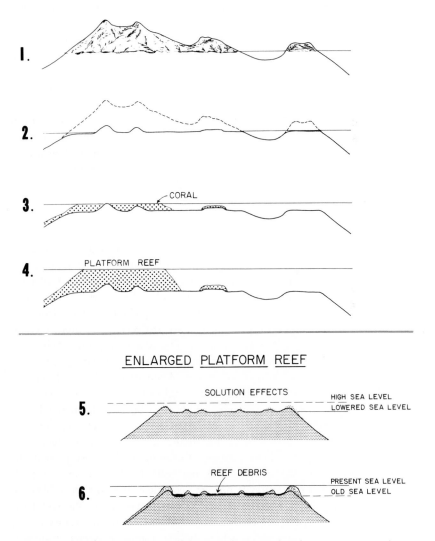

ENLARGED PLATFORM REEF

Figure 108. Development of an atoll by a combination of early wave erosion, subsequent submergence accompanied by upgrowth of the reef, and the effects of solution during the low stands of sea level of the glacial age.

wave-beveled platforms and partly eroded volcanoes (profile 2). These have subsequently sunk. Wherever conditions were favorable for growing reef corals, coral banks and islands kept pace with the slow sinking (profiles 3, 4). In the areas where coral growth was not active, the platforms were left as guyots (see Chap. VIII).

In addition to this long-time subsidence, all of the reefs in their late stages went through an alternate rise and fall of sea level produced by the glacial episodes. This must have given them many of their characteristics. It is not unlikely that the narrowness of the present atoll rims, compared to the wide lagoons, is the result of these glacial stages of low sea level. The temporarily emerged reefs were attacked largely by solution of rain water (profile 5). Such solution was not possible on the growing reefs below sea level because of the lime-saturated nature of ocean water, but it would have started as soon as the reefs emerged. Waves also may have been more effective during the glacial stages because of some killing of the protecting reefs, although most of them probably survived. With a rise of sea level, upward growth along the margins of the temporarily reduced reefs formed the present rims on the edge of a platform consisting of reef material rather than a beveled volcanic island.

Explanation of the Islands

Since the corals do not grow above seal level, some explanation is required for the occurrence of many islands high enough to support a land fauna, including man. The waves are certainly to be credited with the existence of most of these islands because storm waves are known to build ramparts of reef debris up to ten feet or more above sea level. The wind, picking up sand from the beaches, can also build dunes even higher above sea level. The highest elevations of many of the coral islands are dunes. The islands may owe their existence in part to upward movements of the earth's crust, representing temporary reversals of the general subsidence movement. Alternatively, the sea level may have sunk from a higher stand during the postglacial time interval. Some reef surfaces stand above sea level, but most of the reef surfaces from the Pacific atolls are elevated only three to five feet. If these elevated reefs are due to a relatively recent sinking of sea level, we should soon have the evidence, because carbon-14 dating of shells is now proceeding rapidly. At present the carbon-14 evidence, in my opinion, does not seem to favor a time of postglacial high stand, but the record is not very complete. It seems more likely to me that the elevated reefs are due to reversals of movement in the reef base. Similar reversals have been found by Harry Ladd in his recent studies of the preglacial portions of the Eniwetok borings. He found weathered materials, land shells, pollen, and spores as evidence for temporary emergence of the sinking reef masses.

Great Australian Barrier Reef

On the east coast of Australia, north of Brisbane, the continental shelf has a mass of reefs and coral islands extending over a length of about 1,500 miles and a width of about 100 miles. This is by far the largest coral area in the world. The entire shelf is by no means covered with reefs and islands. Actually there are wide channels or passageways, particularly near the mainland, in which there is no evidence of coral bottom. The Australian reefs are still poorly explored, but much information was obtained from the extensive wartime flights of two Australian geologists, Rhodes Fairbridge of Columbia University and Curt Teichert of the University of Kansas. Earlier, the Great Barrier Reef Committee had made various studies of portions of the area and the reports of the biologists C. M. Yonge of Glasgow University and Theodore C. Roughely of the Technological Museum, Sydney, and of the geographer J. A. Steers of Cambridge University are well known. The exploration actually goes back to the eighteenth century when Captain Cook made rather extensive cruises along the Queensland coast. In fact, he had to leave one of his vessels on a reef after it had gone aground during the night and was badly damaged by the rough projecting coral.

The Australian reefs are very diverse in form. They contain many ring-shaped faros, which on a map closely resemble atolls. The profuse coral growth around these reefs is said to be as fine as any in the world. The corals are particularly beautiful and have a great variety of color. They harbor tremendous schools of bright-hued fish as well as other types of organisms, such as the nudibranchs with their swaying purple mantles, the sea anemones, and the many-colored starfish.

Origin of the Australian Barrier Reef

Somewhat the same controversy developed over the origin of the Great Barrier Reef as over that of the atolls. Darwin and his followers thought that these reefs had grown up on a submerging coast. The Daly school, on the other hand, favored erosion of a bench during glacially lowered sea level and post-glacial upgrowth on the outside. Again the controversy has been somewhat settled by borings. The Great Barrier Reef Committee put down two holes, which showed the coral going to depths of 400 and 450 feet. Below this another 200 feet of continentally derived sand and small-sized gravel were encountered. The latter appear to represent shallow-water deposition and indicate, as does the reef material, that sinking has occurred in this area and that the reef upgrowth has been due at least in part to this sinking. Glacial sea-level changes may also have been important. As yet no boring has penetrated the outermost reefs, where the coral may be much thicker.

How to Enjoy a Swim on a Coral Reef

I cannot resist digressing briefly from the primary purpose of the book and giving a little advice on how to enjoy with comparative safety what is to me one of the greatest thrills of life, looking down at a growing coral reef. In various tourist areas there are available glass-bottomed boats, which can give some notion of a reef. Unfortunately, however, most of these boats are available in areas of large population where the water is so much polluted by man that it is not very clear, the corals are stunted (if growing at all), and the fish are scarce. The best way to see a reef is to visit an outlying island and, if you have no scuba gear, swim over it with a face plate, preferably with a snorkel so you do not have to put your head up for air.

Flippers will propel you quietly through the water without disturbing the fish, which are frightened by splashing arm movements, and flippers with solid feet are particularly helpful if you want to stand on the reef. With bare feet it is easy to cut yourself on the sharp coral projections, and coral cuts can be very troublesome, for some of the living coral often gets in such a cut and causes a bad infection. Wearing gloves is also very advisable in order to avoid cuts on the hands and to help propel you over the many very shallow areas where swimming is preferable to walking, because as you swim you can watch the bottom through your mask. If you are walking, it is difficult to see some of the holes and irregularities from the surface. Gloves also will help you to collect corals without cuts and without as much danger from stinging coral, particularly common in the West Indies.

You should be wary of currents. Often the narrow channels through a barrier reef are localities where very rapid currents flow outward, returning the water brought in over the reef by the breakers. These channels are otherwise good places to swim without scraping your knees because they have deeper water, and corals grow profusely on their walls and fish abound. You can usually escape the currents by swimming over to the shallow water on the side, but the currents can be terrifying and can easily sweep you out into the open sea where you would have difficulty and be in danger getting in through the breakers. Wearing flippers is again a safeguard, as they give you great additional speed in fighting currents.

Another danger to consider in reef swimming comes from the fish. Fortunately, except on the outer slopes, reefs are relatively free from the most dangerous sharks. The so-called man-eaters are seen occasionally inside the reefs, but the proximity of very shallow platforms, where sharks are not likely to venture offers some protection. Sharks rarely attack and the attacks are almost always preceded by several investigatory passes by the fish, which can allow the swimmer to get into shoal water where he can stand on the reef. If you see a large shark, it is probably advisable to move rather quietly instead of setting up a great agitation of the water, which sometimes precipitates an attack.

A more common danger on a reef is stepping on a poisonous fish, like the Lion Fish of the western Pacific and the Indian Ocean, the spine-covered echinoids, or the stingrays, which are not always seen when lying on the bottom, particularly when they are partly covered by sand. Similarly, the vicious-looking eels can be a danger to those who are too bold in the exploration of holes in the reefs where these fish are lurking. An attack by an eel except in a hole is almost unknown.

If you use a modicum of caution in your reef swimming, you will probably be far safer than you are in driving on the crowded highways or crossing the streets in a metropolis.

For those of us who love reef swimming, there is another type of danger for the future. Reefs along with other scenic portions of the shallow sea floor are rapidly being ruined by man. This happens in several ways. Corals and beautiful shells are being stripped from many areas by skin divers and scuba divers, leaving only a desert where there was much of beauty. The magnificent fish are being wiped out by the efficient and dangerous spear guns. Finally, as in the case of streams, pollution is destroying the marine life in many places. Why do we not follow the example of the Japanese and set up underwater parks in all of the most scenic areas? To date, we have only four: two in the Virgin Islands, one in the Florida Keys, and one at Point Lobos, California. Pollution control is making some progress but a much more active campaign is needed.

CHAPTER XI

USING THE PRESENT SEA-FLOOR
DEPOSITS TO INTERPRET THE PAST

From Armchair Geology to Rocking Ship

For generations there has been a saying among geologists that "the present is the key to the past," but this has been only a saying, and up to comparatively recent times Mark Twain's remark about the weather, "everybody talks about it but nobody ever does anything about it," was applicable to the situation. As far as the ocean is concerned, the importance to geologists of studying the present lies in the fact that most sedimentary rocks are thought to have been deposited in the ancient seas when they extended over various parts of the present continents. Since the bread and butter of a vast number of geologists is dependent on their interpretation of the conditions under which sedimentary rocks were deposited, it is by no means a waste of time to look at present-day sediments.

The time-honored method of interpreting the sedimentary rocks was primarily philosophical. Using the results of extensive field studies of the rocks, but only a very superficial knowledge of the waves and currents of the present sea and an even scantier knowledge of the sea-floor deposits, the thinkers among the old-time geologists deduced what conditions should be like on the sea floor and what types of deposits should be found in various places. This is a common method and is often referred to as armchair geology. Probably, all in all, the results were fairly good, but as indicated in preceding chapters there were a lot of surprises in store for geologists when the sea was studied more intensively.

Despite the myriad new discoveries resulting from the large-scale oceanographic investigations of recent years, only a small amount of helpful information has been given geologists toward the interpretation of sedimentary rocks. The reason for this deficiency is that most oceanographic studies have been concerned with the deep oceans, and unless the common opinion of geologists is a mistaken one, the majority of the sedimentary rocks were deposited in shallow seas. On a smaller scale the oceanographic institutions and various laboratories have made a beginning in the study of shallow-water sediments. One geologist, P. D. Trask late of the University of California, collected scattered samples from various parts of the world in a search for sources of petroleum. When it was compatible with the rest of the program, geologists have often accompanied fishery research vessels in shelf cruises in order to gather a little information. Beach and shore-line studies have also become important factors in recent years. All of this has helped scientists to understand the conditions of deposition of sedimentary rocks.[1]

Perhaps the largest-scale attempt to study recent sediments as a key to the past has come from a project that was initiated by the American Petroleum Institute in 1951. This project was instigated largely by Shepard Lowman, who was among the first oil geologists to appreciate the importance of knowing more about present-day deposits. Committees of oil geologists spent a number of years trying to decide the best means of remedying the admitted ignorance of geologists about the conditions of deposition of sedimentary formations similar to those in which oil is found. Finally an agreement was reached, and a program of study was established and given to the University of California to administer. This was conducted largely by Scripps Institution scientists, despite the fact that the first field of study was the shallow waters of the northwest Gulf of Mexico. The second field study was in the Gulf of California, work that was completed in 1963. The results of these studies have provided some helpful clues, which are already being used to some extent in interpreting old sediments.

Origin of Sandstones

To show how knowledge acquired from the study of recent marine sediments applies to rocks, we can give some examples of interpreting common types of sedimentary rocks with the help of knowledge of present-day conditions. Sandstones are composed of grains of sediment that are visible to the naked eye, ranging from 1/16 millimeter to 2 millimeters in diameter. These sedimentary rocks originate as either land or sea deposits. In many cases there is no good way to tell which origin they had, especially since fossil organisms are apt to be rare or missing in sandstones. Several types of sandstones appear to be either marine sediments or deposits such as barrier islands, which are on the borderline between marine and continental sediments. These marine and borderline types will be considered here.

Sheet sands. Sandstones can be traced for scores or even hundreds of miles. Many of them have been interpreted as river deposits, formed by braided streams spreading over the plains at the base of the mountains. Others, however, have marine fossils, and these offer some difficulties of explanation. Sands are common on the continental shelves, but most of them appear not to have any source from the lands at the present time. Where sediments are entering the sea, as at the mouths of rivers or at the base of rapidly retreating sea cliffs, sand is usually confined to a narrow belt along the shore, and muddy sediments occur outside. The broad offshore sand zones are generally found beyond this mud (Fig. 109) and hence have been interpreted as being a relic of earlier

---

[1] An excellent summary of these investigations is given in *Principles of Stratigraphy* (New York: John Wiley & Sons, 1957) by Carl O. Dunbar and John Rodgers.

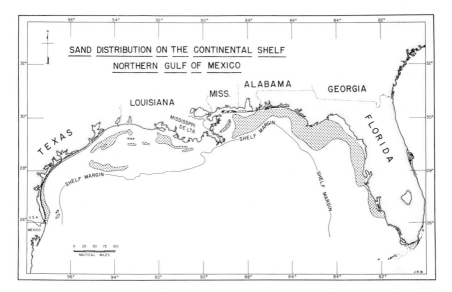

Figure 109. Showing the somewhat haphazard distribution of sand deposits on the continental shelf of the northern Gulf of Mexico. Most of the remaining area is covered with mud.

conditions when the sea level was lower. Thus they may represent stream deposits not yet covered or old beaches formed successively inland during a rising sea level.

Some of the marine sheet sands, however, do not appear to be either beach or river deposits, since their faunas are strikingly like those of the open sea. An explanation for these sheet sands may come from the studies along the continental shelf of the Gulf of Mexico. There, sand occurs in broad belts, some of it along the outer shelf. Most of it seems to have no present source from the mainland because the rivers are now depositing their sands in the heads of bays or as bars at river mouths, as off the Mississippi, Brazos, and Rio Grande rivers, which empty into the open Gulf. The shelf sands have faunas that are largely marine, although occasionally old estuarine shells are found within them. In some places foraminifera faunas characteristic of the present depth of the water are found as much as several feet below the bottom mixed with shells that lived at much shallower depths. Since the sands are not now being supplied, it seems reasonable that old sands and shells are being moved around the ocean bottom by present-day currents and the present-day foraminifera are accumulating in them. Fossil remains are scarce in river deposits. Therefore, when the seas covered the area, the faunas living during the reworking period would have been the ones preserved in the sediments. This reworking by the sea of older nonmarine sands may explain many marine sandstones, particularly where the sandstones are widespread but not very thick. During the past there seem to have been many invasions of the sea that would have produced conditions favorable for this reworking.

Another cause of marine sheet sands may be powerful tidal currents. As an example, the strong tides that run along the outer arm of Cape Cod are carrying the sands eroded from the sea cliffs of the Cape both to the south and to the north. The largest deposits of these sands are forming in what is called Nantucket Shoals, south of the Cape. These shoal areas of shifting sand could easily be preserved as a rather irregular sheet of sandstone. Similarly, the southern end of the North Sea is predominantly sandy and receives its source of sand largely from the strong tidal currents coming up through the narrows of the English Channel. Here again the sands are irregular and shifting, but a general emergence of the area would yield a marine sandstone that would contain a few remnants of land and beach deposits, such as the Dogger Banks from which evidences of fossil man have been dredged. A. H. Stride of the British National Institute of Oceanography has discussed the extensive sand plains along the continental shelf southwest of the English Channel. Many of these have been found to have great sand waves running across them. This has been determined by an instrument with a new acoustic method of side scanning the bottom. Tidal currents appear to operate effectively, even on portions of the outer shelf in this area. The broad sand flat at Mont-Saint-Michel on the Normandy coast is essentially a beach deposit because this seven-mile flat is alternately covered and uncovered by the tide. Nevertheless these deposits should resemble sheet sands if preserved as rock formations.

Oil-rich sandstone stringers. Other sandstones occur as elongate lenses (called *stringers* or *shoestring sands*) rather than as sheets. These stringers are of great interest because they contain much of the oil wealth of the world. Many of them are very probably river sands, resulting from the filling of stream channels, but others have been interpreted as old barrier islands similar to the great sand islands that extend along the Texas coast. Once oil is found in a stringer, it is very important for oil geologists to know which origin is involved, so that they can predict the trend and shape of the sand body. The studies of recent sediments have provided some clues that should be helpful in distinguishing between these two types of elongate sand bodies. The difference is striking.

It is easy for American geologists to learn the general geography of barrier islands because barriers are almost continuous from New York Harbor down the east coast and around the coast of the Gulf of Mexico to a point well south of the Mexican border. In fact, they follow the coastal lowlands of the United States just as they fringe such lowlands in other portions of the world. Almost all of them have a straight or gently curving side toward the ocean, whereas the side toward the lagoon or marshland is usually scalloped (Fig. 110). The reason for this difference is that the ocean waves straighten the exposed side,

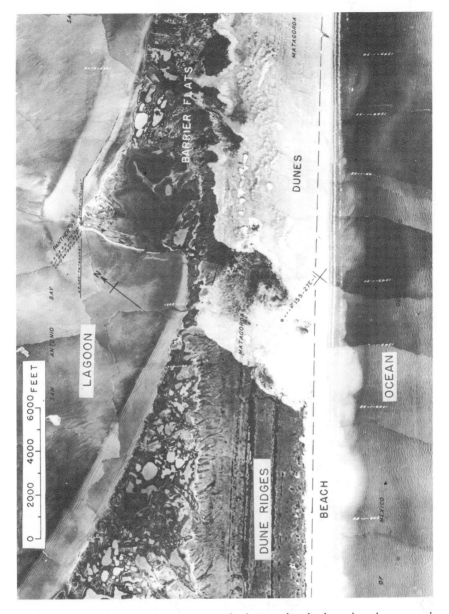

Figure 110. Aerial view of a typical barrier island. Note that the dunes have been covered by vegetation where dune ridges are indicated, but elsewhere the dunes have been spread widely over the island and are not controlled by vegetation. The ponds in the barrier flats are unusually numerous. Dashed-in black line added to indicate inner margin of beach. Photograph is an aerial mosaic by Edgar Tobin Surveys.

and fans are built on the inside by washing over of waves or by the growth of tidal deltas inside the migrating inlets.

The stream channel sands, on the other hand, usually have sinuous boundaries on both sides and the whole channel is winding, following the general sinuosity of a river crossing a lowland. Furthermore, the average river trends approximately at right angles to the coast, whereas the barrier islands run roughly parallel to the coast. This can be important where the general relations of the land and sea at a particular time are known.

In cross section, the barrier is usually broader at the base and narrower toward the top (Fig. 111, Sec. A), but it is possible for the top to migrate landward across the base, so that the barrier becomes lopsided. Our studies have indicated both that the barriers may migrate landward and that they may build seaward. A period of stormy weather will tend to drive them landward, whereas calm seas are a time when they build toward the ocean. A river-channel filling (Fig. 111, Sec. B) differs in having a narrow base and a broad, relatively flat top in contrast to a curving top for a barrier.

The detailed structure shown in a cross section of a barrier (Fig. 111, Sec. C) is also quite different from that of a river-channel deposit. The barrier has the foreshore sloping beds of the beach along the ocean front. Next are the horizontal beds of the beach berm and then ordinarily a mass of dunes with cross-bedding of diverse character. On the lagoon side, the washover fans have gently sloping beds with an inclination toward the lagoon. The river-channel deposit, on the other hand, often has beds dipping away from the margins along both sides, caused by slumping, and a series of lenses in the middle of the section (Fig. 111, Sec. B).

The sediments are also quite different. The barrier has well-sorted sand in both the beach and dune, the sand in the latter usually being somewhat finer and more rounded. The washover fans often contain muddy material deposited with the sand because the fans have an abundance of ponds into which dust is blown. Pollen of plants and products of evaporation of the ponds are also common on the flats. The river sands are more poorly sorted and have many zones of mud along with gravel lenses interspersed with them. The same poorly sorted material extends across the entire stream valley (Fig. 111, Sec. B) because the channel migrates from one side to the other.

Studies of grain orientation show another striking difference. The river sand grains show an elongation parallel to the general direction of the channel, whereas the grains in beaches are commonly elongate at right angles to the shore, and the same would be true in the washover fans. The dunes, however, will have only weak preferred orientation, with some tendency of the grains to line up with the prevailing wind.

Fossil faunas will often tell the difference between barriers and stream deposits. The barriers have an abundance of marine shells on the ocean side, many of the shells being carried also into the dunes. Lagoon shells may be found

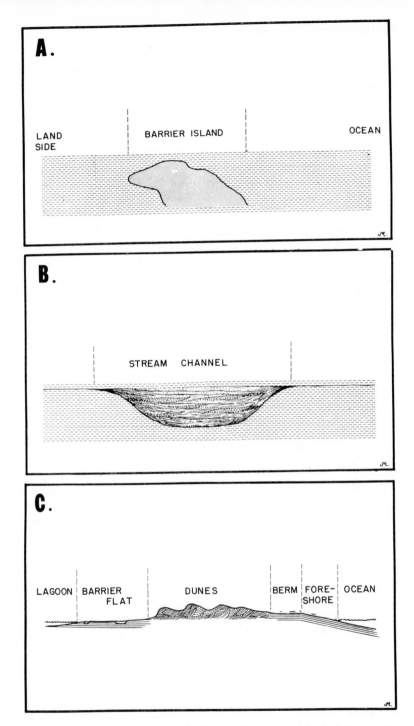

Figure 111. Sec. A: Showing how a barrier-island deposit might look in a section of sedimentary rocks. Sec. B: Cross section of a typical stream channel fill with lenses of coarser sediments representing the shifting stream patterns. Sec. C: Cross section of a barrier island showing the inclination of the bedding planes and the cross-bedding of the dunes.

on the inner margin mixed with some ocean shells. The river-channel deposit will have few fossils, if any, these being of the fresh-water type except where the river has reworked old marine deposits and re-incorporated them with the river deposits.

The importance of barrier islands as sand traps for oil has often been overlooked. Most of us who were trained as geologists some years ago were taught that barriers are ephemeral features formed on coasts of emergence, where they would be eroded by rivers by further emergence or, if the coast remained stable, would be pushed shoreward by the waves, first filling the lagoon and finally being washed away entirely by the waves. Our recent studies have shown that the barriers along the coast of the Gulf of Mexico were built, at least to considerable extent, as the sea was rising because of the melting ice at the end of the glacial period. This represents submergence rather than emergence. Barriers also formed around the margins of deltas where deposition had stopped and submergence occurred because of the compaction of the deltaic sediments. Quite independent of this deltaic sinking, there has been more or less continuous submergence for millions of years along many coasts, such as Texas and Louisiana, carrying down barrier islands of various ages so that these islands have been preserved to act as reservoirs when the oil migrates into them from adjacent muddy marine formations.

Deep-sea sands and their recognition. Where the sandstones were found as thin distinct layers between thick masses of shale, the latter having every indication of deposition in quiet water, the geologists have often been at a loss to explain the sudden changes that introduced the sand. Many of these sands are now interpreted as being the result of turbidity currents. These deposits appear to have been formed in relatively deep water, perhaps in some cases at true abyssal depths of the ocean floor. Until turbidity currents, first confirmed by experiments (Chap. I), were suggested as the explanation for deep-sea sand layers at many localities around the continental margins, all sandstones were interpreted as being of shallow-water origin, not only because the sand required active currents but also because the ripple marks and cross-bedding contained in these sands were thought to be developed only in shallow water. Now it is clear that turbidity currents can produce all of these effects even in abyssal depths. Ordinary currents, however, now known to exist in deep water, may have been important in addition to turbidity currents.

Relative to these sands, an interesting event in the history of geological thought resulted from the work of Manley Natland[2] during his student days. Natland decided it was time that someone investigated the foraminifera in the deep basins along the California coast. Taking a very small boat, and using very primitive homemade equipment, he sampled the channel between Long Beach and Catalina Island. Studying the foraminifera, he found that there were

---

[2] Now Chief Paleontologist for Atlantic-Richfield Oil Company, western division.

definite depth-zone assemblages. Applying his newly acquired knowledge to a land section near Ventura, California, he decided on the basis of fossil foraminifera that some of the rocks had been deposited in water several thousand feet deep.

When Natland's report first appeared, it caused little but scorn among the California geologists because they noted that the supposed deep-water formations contained sandstones and even conglomerates and that the sandstones had ripple marks and other indications of strong currents. At Scripps Institution we began to check Natland's faunal zones by giving him our deep and intermediate-depth samples and asking him to tell us the depth of deposition. His success surprised us. Despite heavy opposition Natland held his ground, but he was somewhat at a loss to explain the presence of the coarse sediments. Finally, Kuenen came to his rescue with the turbidity-current tank experiments, and the oceanographic institutions began to find deep-sea sands. The battle was won and only a few ultraconservatives are still holding out for shallow origin of these coarse sediments.

The extensive publicity that has come from the turbidity-current discoveries may have caused some geologists to use them as a catchall. One keeps hearing of new cases of turbidity-current deposits among the sedimentary rocks. In fact, it is getting to be a means of explaining almost any anomalous situation. We still do not know enough about the deep-sea sands to be sure of their identification in ancient rocks. A common criterion that is used is called *graded bedding,* that is, coarse sediments below grading up into fine. This grading was found in the deposits resulting from Kuenen's experiments. However, graded bedding is found also in some stream deposits and may very well occur in the sea where a storm has stirred up the bottom sediments and they sink back with the larger grains beneath. Furthermore, conditions were probably not very good for developing turbidity currents in many of the seas of the past because, as far as we know, in order to develop a turbidity current it is necessary to have a relatively long slope on which landslides can take place, becoming turbidity currents as they take on water in moving downward. This means that a deep basin is necessary for turbidity-current deposits.

In a deep basin the normal sediments that constitute the deposition between turbidity-current flows should have many characteristics indicative of deep-water origin, including their faunas. If foraminifera are not available or if the deposit is too old to allow interpretation from present-day depth-zone assemblages, there are other means. For example, a shallow-water origin is indicated by finding fossils indicative of shallow water in the sands between the shales. Many of these forms required light and photosynthesis, which is confined to shallow water. The absence of the shallow forms is not significant unless it is accompanied by the presence of remains of deep-water organisms. Fish fossils may be very helpful because some fish, of the scavenger type, live only in deep water. Glass sponges, stalked crinoids, the cephalopod Nautilus, and many of the holothurians (sea cucumbers) are fairly reliable indicators of deep water at present but cannot be used in periods much older than the Tertiary. Also a fauna consisting largely of brittle stars is often indicative of considerable depth of deposition. As far as we know, many of the types of shells are confined to shallow water. Unless these are carried down by a turbidity current or a landslide and hence are found in or just above the sand layers, they can be used successfully to distinguish shallow water origin.

Shales and Mudstones

The sedimentary rocks, which are made up dominantly of grains of silt and clay, are called *shale* if fissile or laminated, or *mudstone* if layering is rare or well separated, that is, by several inches. *Siltstone* is a name applied to rock known to consist largely of silt-sized particles, and *claystone* to rock in which the constituents are clay-sized or consist mostly of the common clay minerals, illite, kaolinite, and montmorillonite (determined principally by X-ray diffraction techniques). In the present discussion little attempt will be made to differentiate the types of mud rock. Shale, which is the popular name, will be used somewhat indiscriminantly for all mud rocks.

An interesting thing about shale is that it probably constitutes more than half of all the known sedimentary rocks, and sandstone constitutes only about 20 per cent, whereas more than half the present-day sediments on the continental shelves are sand and only about 30 per cent are muds. Yet most observed shales are thought to have been formed in shallow water with depths comparable to those of the shelves.

The discrepancy between sedimentary rocks and the shelf sediments perhaps can be explained by the fact that at present there are few inland seas, partly because the sea level is now lower than it has been in most past ages. If the ice of Antarctica and Greenland melted, returning water to the ocean, the sea level would be raised about 200 feet. This ocean rise would submerge valleys and lowlands and produce many protected bays and seas where, because of less wave and current action, the muds would be deposited more readily.

Shelf mud deposits. As far as we know, the areas where active deposition is now occurring on the continental shelf are largely covered with mud that will form shale if it is buried and turned to rock. In the well-studied shelf off Texas and Louisiana rather extensive areas are receiving mud deposits at an appreciable rate, mostly derived from the Mississippi or other large rivers entering the ocean. It is a common thing on the outer shelf to find that some fifteen feet of such deposits have accumulated in the past 10 to 15 thousand years since the ocean spread over the area. These deposits have little stratification, apparently because the bottom fauna is constantly churning over the sediment and destroying the layering. Nearer the shore, thin discontinuous laminae of sand are found interlayered with the mud. On the outer shelf, the foraminifera are largely planktonic, floating around in the water and sinking to the bottom after death. Nearer shore, on the other hand, the foraminifera in the sediments are largely benthonic, or bottom-living. In ancient sediments this difference between inner- and outer-shelf foraminifera helps to establish the locus of deposition, although it is not an invariable rule. Many of the Tertiary mudstones found in oil wells along the coast of the Gulf of Mexico appear to have had a shelf origin, as Shepard Lowman first demonstrated from the foraminifera.

Lagoonal muds. The extensive lagoons or bays inside barrier islands have accumulating in them mud that is different in many respects from the mud of the open shelf. It should be possible to distinguish the shales of the two environments in most cases. This is important in looking for oil, since it indicates which way to move in order to get into the oil-rich barrier-island environment. The barrier should exist gulfward if the shale formation was lagoonal, and toward the continent if the shale was open shelf in origin.

The faunal differences between gulf and lagoon are pronounced. The variable salinity in typical lagoons, compared to the rather constant salinity on the outer shelf, means that the lagoonal animals and plants have had to be capable of tolerating great changes. Temperatures also have wider annual ranges in the lagoons, more heating in summer and more cooling in winter. The result is that only a few species have withstood this variability. The few that do live are apt to be very abundant, whereas in the open sea there may be fewer total numbers of individuals but many more species. Some groups such as echinoids do not get into the lagoons to any extent but are common on the continental shelf. On the other hand, some species, notably oysters, are largely confined to the lagoons, where there is low to intermediate salinity. Thick oyster and mussel reefs may form there. These are very distinctive where they cut across shale formations.

The lagoon deposits rarely contain glauconite, whereas it is quite common on the continental shelf, particularly if deposition on the shelf has been slow. However, in our experience faecal pellets are more often seen in the lagoon muds than in those of the open shelf.

Stratification does not seem to be more common in the bays than on the outer shelf. The organisms living on the bottom in this environment are equally active. An exception occurs at localities where streams are forming deltas, but the same exception is found on the continental shelf near where deltas are being built. Another exception is found in some of the shallow lagoons in semiarid regions. Because of the very high salinities and stagnant conditions here, bottom-dwelling organisms are limited and hence the stratification is preserved. These same dry-area lagoons are apt to have sand grains coated with calcium carbonate, sometimes forming the rounded shapes called *ooliths* or *oolites,* which are more common in limestones. Gypsum is found elsewhere in these arid regions, and salt deposits may develop on a large scale if evaporation is excessive and restricted inflow from the open sea is maintained.

Delta margin muds. A curious enigma exists relative to deltas. There seems to be no doubt that the bulk of the sediment that is now being carried into the open ocean and into bays is being deposited largely at the river mouths as muds, which form the submarine portion of deltas. Shelf deposition rates are commonly as much as a hundred times as fast near the river mouths as they are at a distance. There is no reason to doubt that this was also true during the past. Yet until recently the literature on sedimentary rocks contained only a few references to ancient deltas. Also, instead of consisting of shales, most of the recognized fossil deltas are largely sandstone, as, for example, the sandstones of the Catskill Mountains. These are interpreted as an ancient delta built from Appalachia, a former elevated land mass to the east that has now disappeared. Since the deposits from the present-day large rivers are predominantly muds, where are the delta-front muds among ancient sediments?

Some of our studies have led us to wonder if many ancient deltas have been overlooked among the ancient sedimentary rocks. Such an oversight might have occurred partly because geologists have been looking for steep foreset beds along the front of an advancing delta (Fig. 112). Actually, the Mississippi Delta and most other large deltas are building forward on very gentle slopes of less than one degree. With a very slight but perceptible discordance in angle between the topset and the foreset beds, the old deltas could easily be misinterpreted unless other criteria were used.

Among the features that serve to distinguish delta-front deposits from other continental-shelf or bay sediments is lamination, which is generally not destroyed in the delta deposits because the fresh water inhibits the development of the bottom dwellers that destroy lamination. Also, deposition is so fast around most deltas that even if the organisms did exist, layering would probably not be obliterated before the sediment was deeply buried. Deltaic sediments usually contain a large amount of wood fragments introduced by the streams. Mica is apt to be much more common than in nondeltaic shelf deposits. Fossils other than wood will be very scarce, and like the deposits in the lagoons the species will be very limited in number due to the changing conditions of salinity and temperature. Because the rivers have shifting mouths and old abandoned delta lobes sink, due to compaction of their muds, the sediments of a delta are apt to show many alternations between marine and land conditions. These alternations may be cyclic, such as those referred to as cyclothems, which have been found in many places in the deposits of the great coal age called the Carboniferous, which includes the Mississippian and Pennsylvanian periods. The importance of deltaic deposition among the shales of the Pennsylvanian period has been stressed recently by H. R. Wanless of the University of Illinois. Such cyclic deposits developed also in the Pleistocene as the result of swinging sea levels, and it may be that similar changes of sea

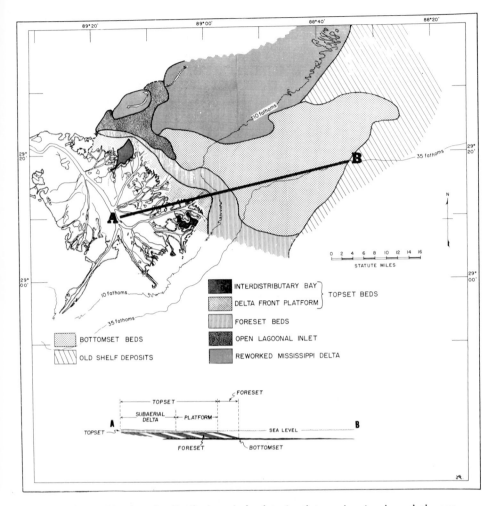

Figure 112. Showing the distribution of the deposits that are forming beyond the eastern front of the Mississippi Delta. The small diagram at the bottom shows how these deposits fit the geological concept of topset, foreset, and bottomset beds except that the foresets off the Mississippi Delta, unlike the diagram, have an inclination of only one-half of one degree. The bottomset beds overlap onto old shelf deposits to the east and onto reworked older Mississippi Delta deposits to the north.

level occurred during the coal periods. Conditions that may be somewhat similar to the swamps in which coal was formed are found around many deltas of the present day, including the lower Mississippi Delta with its extensive marshes in which peat is forming. When deeply buried, peat turns into coal.

Black shales. A problem that has caused great discussion among geologists is the origin of black shales. The Chattanooga shale of the South and Midwest is particularly puzzling. In addition to its dark color it is well laminated and has few bottom-dwelling organic remains, suggesting that bottom organisms did not burrow into it. The fossils consist of animals and plants that could have sunk from the surface water. The lack of bottom-dwelling organisms and the black color may indicate stagnant conditions. Oxygen is soon exhausted under these conditions, and the plants are preserved as a black muck, which permeates the rest of the sediments, giving them a dark color. Some investigators have thought that these ancient seas were deep like the Black Sea and the basins of the Norwegian fiords. Here the bottom water is stagnant due to lack of circulation. Black muds are forming in some fiords, but the muds are more gray than black in the Black Sea despite the stagnation. However, the Chattanooga shales directly overlie an erosion surface and are covered by limestone believed to be of shallow-water origin, so that the case for deep water is not well substantiated. W. H. Twenhofel suggested that luxurious growth of seaweed in very shallow water might sufficiently hamper the waves and so allow the black muds to accumulate and stop the circulation necessary for existence of most marine and lagoonal animals. Black muds are found in part of the Laguna Madre of southern Texas, where lack of oxygen exists under an algal mat.

The black color in shales does not necessarily mean stagnation, because the color may form as the result of a great abundance of organic matter that overbalances the loss of organic matter from decay on the bottom or from feeding by other organisms. In such cases there should be an abundance of organisms and the lamination should be considerably disturbed. This is true of some of the shales of the German Rhineland and in the modern counterparts found in Danzig Bay and elsewhere in the Baltic Sea.

Deep-basin muds. The deposits of the deep basins and troughs off California are primarily mud, although they are interbedded with sand layers of presumed turbidity-current origin. The color of these muds is generally dark green, although there is a fairly high content of organic material. Here, weak circulation prevents complete stagnation, and the muds are not very well stratified. Counterparts of the deep-basin muds are found in the Tertiary sediments of southern California, particularly in the Los Angeles and Ventura areas, where thousands of feet of sediments accumulated under former deep-basin conditions. The depth of water is recognized from the foraminifera and fossil fish. These old basins are well known because much of the oil wealth of California is obtained from their sediments.

Another type of deep mud deposit has been recognized recently as the result of work by David Ericson of Lamont and K. O. Emery of Woods Hole. They found that rather thick, gray silty muds occur in basins that contrast with the finer clayey sediments on either side in having few foraminifera and little burrow mottling, except in their uppermost portions. These they ascribe to deposition by the tail of large turbidity currents that were depositing more sandy sediments in shoaler areas.

Limestones and Dolomites

Sedimentary rocks that consist of more than 50 per cent of the minerals calcite and dolomite are called *limestone* if calcite is the dominant mineral, and *dolomite* or *dolostone* if the mineral dolomite is more abundant. If these lime rocks are fine grained and so poorly consolidated that material can be rubbed off, they are called *chalk,* or if they are high in clay content, *marl.* These lime rocks are about as abundant as sandstone, but they are not so common as shale. Their origin has been determined by comparison with recent lime sediments, most of which occur in tropical areas. The fact that limestones are so abundant in areas now having temperate and even frigid climates seems to indicate generally warmer conditions for these areas during most ancient periods.

Fragmental limestones. On many continental shelves in the tropics and even in the subtropical areas the sediments consist principally of broken shells, coral fragments, or other hard parts of organisms. The studies by Howard Gould and Robert Stewart for the United States Geological Survey showed that such fragmental materials were widespread off western Florida. The wide bank off Yucatan is also said to have the same materials. Around various tropical islands, such as Samoa, and in the Java Sea detrital lime deposits cover most of the bottom. These sediments have been transported by circulating currents and wave action that spread the fragments of organisms along the bottom. Hence, it is called a fragmental limestone. In all tropical areas that lack a good source of land-derived sediments, the lime sediment forms the principal deposit. When it is cemented together, it becomes a limestone.

Careful studies of European limestones have shown that they are predominantly fragmental. We do not know as much about American limestones, but it seems very likely that they are also more of the fragmental than of other types. One can recognize fragmental limestone by the scattered and broken shells that are found abundantly through the formation. When a limestone has been converted to dolomite,[3] the shell structures are often destroyed so that fragmental dolomite is very hard to recognize.

Reef limestones. Modern reefs have been described in the chapter on coral reefs (Chap. X). Similar reefs were found in the seas of the past and many have been preserved as sedimentary rocks. As a source of petroleum these ancient

---

[3] It was formerly believed that all dolomite was originally deposited as limestone and subsequently altered to dolomite. Now we know that dolomite is deposited directly in a few saline lakes and in some coastal lagoons of high salinity.

reefs are fully as important as the stringer sands. The cavernous nature of coral reefs has made them particularly good reservoirs for the oil that has migrated into them from the surrounding formations.

Ancient reefs are not always easy to recognize, partly because many of them have turned to dolomite and the structure of the corals and other reef-builders has been eliminated in the process. Reefs are often distinguished by the coral talus or debris that was broken from the reef front and accumulated on the slope next to the reef. Often these fragments contain better-preserved corals than the reef itself. An interesting example of an ancient reef was found in rocks in western Texas (Fig. 113). Its interpretation by Philip King of the

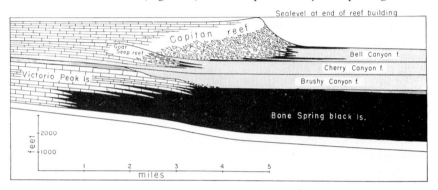

Figure 113. The diagrammatic cross section of El Capitan reef in west Texas. This ancient reef had a slope below it like modern reefs. From P. B. King, U.S. Geological Survey.

United States Geological Survey shows a great reef (part of which now forms El Capitan, the highest mountain in Texas) bordered by a large talus slope out in front, but the talus is now solid rock. A deep basin, partly filled by sand washed over the reef, existed east of the reef. West of the reef there was a shallower lagoonal basin bordered by an arid land mass. Because of the excessive evaporation in the lagoon, gypsum deposits developed there except near the reef, where fresh sea water was washed over into the lagoon. The reef apparently consisted largely of algae and sponges rather than coral.

Another example of an ancient reef is now found among some of the most beautiful scenery in the Alps. The massive vertical-sided mountains of the Dolomites in Italy and Austria with their many pinnacles are largely structureless dolomitic rock. They are believed to have had their origin as coral reefs. Some of the corals are still preserved, and the lack of stratification in a mass of lime rock is hard to explain in any other way.

Chalk and fine-grained limestones. Chalk is the weakly cohesive lime rock that was used for generations for writing on blackboards.[4] Microscopic exami-

---

[4] A synthetic chalk consisting largely of gypsum is now used.

nation showed that chalk contained an abundance of the free-floating plank-tonic foraminifera of which Globigerina is particularly common. It was thought for a long time that chalk was a deep-sea deposit, being compared to the globigerina ooze collected by the *Challenger* Expedition in the last century. However, more careful study of chalk formations, such as those of the cliffs of Dover, showed that associated with the foraminifera there are many other organisms that are characteristic of shallow water. The deep-sea origin has therefore been abandoned. Some of the chalks of our southern states have been shown by M. N. Bramlette, of Scripps Institution, and others to have an abundance of the minute lime-secreting marine plants known as coccoliths. Bramlette considers that other chalks, such as that of England, may also have formed as a deposit of coccoliths, but that the coccoliths have largely disap-peared under pressure, being altered into small calcite particles.

Some of the chalk formations may turn into fine-grained limestones when deeply buried. In Tunisia, the chalk formations in an unfolded area can be traced to a folded area where the chalk has turned to a fine-grained limestone. Some of it may represent coccolith deposits. Elsewhere, lime mud like that of the Bahama banks has very probably been caused by chemical precipitation from the supersaturated water. This mud may also turn into a fine-grained limestone.

Oölitic limestone and the Bahamas. Many ancient limestones have been found to contain a mass of small round grains looking like fish roe. This ap-pearance has led to the name *oölite,* from the Greek *oö,* a shortening of *oion,* meaning egg. In cross section the round grains show concentric rings. The Jurassic rocks of England were first called the *oölite series* because of their high content of *oölitic limestone.* This name, introduced by William Smith about the year 1800, is now used only for more restricted formations.

Modern oölites are forming in various shallow warm-water seas, all having high or relatively high salinity. The oölites have been studied particularly in the Bahamas, where they form spectacular underwater dunes, called locally *bores.* These are plainly visible in flying over the shallow banks. These dunes are shaped by the waves and currents and are in fact an excellent indication of the dominant currents in the different parts of the area. If these are later turned to rock, a cross-bedded oölitic limestone will form.

The investigations of the English geologist L. V. Illing and of Norman Newell of the American Museum of Natural History and Columbia University have been particularly fruitful in explaining the Bahama oölites. The oölites evidently are formed by chemical precipitation from the cold deep waters warmed by rising onto the banks and becoming supersaturated. Deposition occurs, especially near the edge of the shallow platforms and at the edges of tidal channels. Various nuclei are coated with layers of calcium carbonate consisting of the mineral *aragonite.* The grains are rounded partly by deposi-tion and partly by abrasion from rolling around in the shallow water. In the areas where these underwater dunes of oölite occur there are very few bottom-living organisms except echinoids, probably because of the lack of bottom stability. Elevated Pleistocene limestones of the Bahamas are commonly oölitic also. These oölitic limestones can be seen all along the ridges of the island Eleuthera, perhaps the most beautiful of the Bahamas.

Oölites are found in a wide belt on the outer shelf off western Florida. They are said to represent old deposits formed when the sea level was lower during the glacial stages. They are forming today in the supersaline Laguna Madre along the southern Texas coast. Here they form only adjacent to the shore where the small lagoonal waves are breaking. These oölites are lime coverings of quartz sand grains, in contrast to Bahama bank oölites, which either are faecal pellets or are composed entirely of calcium carbonate. Oölites are not all marine. They form in various salt lakes, notably Great Salt Lake, where they have been studied by A. J. Eardley of the University of Utah.

Deep-water limestones due to turbidity currents. Investigations of the sediments in the deep troughs between the Bahama Islands showed the Lamont scientists and G. A. Rusnak of the U. S. Geological Survey that masses of the shallow-water lime deposits of the Bahama reefs have been transported down into the deep basins. Apparently this deposition has been by turbidity currents. The materials that make up these deep-water layers are very similar to those of the shallow water, but some deep-water foraminifera are found within them.

Puzzling Conglomerates

Conglomerates are rocks containing rounded pebbles, cobbles, or boulders in a matrix of finer material. Most conglomerates are stream deposits formed principally on land. It is possible for a river to carry pebbles into the sea, or pebbles may be distributed along the shore by longshore currents coming from wave-eroded beaches. Generally such conglomerates are limited to a narrow fringe and would not be difficult to recognize in rock formations because of their linear arrangement.

Conglomerates that have puzzled many geologists and led to all sorts of explanations are those in which pebbles are scattered rather heterogeneously, mostly in marine formations. The matrix of such conglomerates usually shows no relationship to the pebbles and may consist of very fine muds or even lime deposits.

Study of recent sediments has helped explain some of these unusual conglomerates. In studying the cores taken by Charles Piggot across the Atlantic in 1936, M. N. Bramlette of Scripps Institution and W. H. Bradley of the United States Geological Survey were impressed by layers with many pebbles, which alternated with normal deep-sea deposits. They concluded that the pebble layers were ice-rafted by the abundant icebergs of the great continental glaciers that were drifting through the North Atlantic during glacial epochs. Similar pebble beds are found in the deposits around Antarctica, where icebergs abound today. Somewhat more puzzling are the scattered pebbles found in areas like the Gulf of Maine and the Gulf of St. Lawrence. These pebbles are not confined to layers below the bottom but are found in grab samples taken right at the bottom. They may represent iceberg deposits that have not been completely covered since the glaciers retreated some 10,000 years ago. Pebbles in the bottom sediments off northern Alaska have been attributed to the summertime breaking away of the coastal ice that has frozen onto the margin of the land and picked up pebbles from the shore. The ice drifts away from shore and when it melts, drops the pebbles.

Conglomerates formed by ice drifting are probably not very common among sedimentary rocks older than the Pleistocene, largely because the climate during past ages rarely got cold enough to form glaciers or shore ice. However, there seems to be little doubt that there were large glaciers about 200 million years ago in the Permian period and perhaps also during the Carboniferous (coal period), which preceded it. Less positive evidence has been used for glaciation at even more remote periods, including the pre-Cambrian, more than 500 million years ago. Some of the conglomerates are most easily explained by ice-rafting during these times of cold climate.

Pebbles are also introduced into the sea by various other means. Sea lions pick up pebbles from the shallow ocean floor and carry them in their stomachs, apparently to aid digestion. A former assistant of mine, Kenneth O. Emery, now a well-known marine geologist of Woods Hole Oceanographic Institution, used to go up and down the beach in front of Scripps Institution looking for dead sea lions. When he found one, he would horrify all the people on the beach by cutting open its stomach and looking for pebbles. He found a good many during a sea-lion epidemic, but the sea lions cannot be expected to be a large source of conglomerates. There are not enough of them and probably never have been. Kelp, which grows on rocky areas along many coasts, has holdfasts, very much like roots, with which it hangs onto the bottom. The plant has a long stem, sometimes a hundred feet long, and at the top, floating on the surface of the water, there are broad fronds with bulbous floats containing a gas. When the kelp is broken off by storms or when the plant dies, the holdfast may bring along rocks from the bottom, and the drifting plant held up by the bulbs may carry these rocks for a considerable distance. Kelp is thus capable of transporting pebbles and eventually, when it decays or is eaten, of dropping them to the bottom. Unless kelp was much more abundant in the past, this also is probably a minor source.

The drifting out to sea of mats or rafts of floating land vegetation is a well-known phenomenon in tropical areas and may lead to the introduction of many pebbles, carried in tree roots, particularly around the mouths of rivers. Some of the scattered pebbles in deltaic formations may be explained in this way.

Probably turbidity currents carry pebbles out along the course of submarine canyons and even out onto the deep-sea fans beyond. We first found such pebbles in a core taken from a deep portion of Monterey Submarine Canyon. Maurice Ewing and others from Lamont Geological Observatory later reported many other cases from the Atlantic and Caribbean. These currents may be an important source of conglomerates in ancient rocks. For example, the cobble beds in the mountains near Ventura, California, have been attributed to either turbidity currents or submarine landslides into the deep trough that existed in this area during the Pliocene. Many of the conglomerates of the Alps were formed in the same way during the Miocene. Probably other causes of marine conglomerates will be discovered as studies progress.

It is not always possible to tell the environment in which a sedimentary rock was deposited. A special problem is found where the rocks have been subject to considerable heat and pressure, both of which may change their characteristics to such an extent that they lose the earmarks of their depositional basin. As we acquire more familiarity with present-day sediments, many more ways should be found in which the origin of the sedimentary rocks can be diagnosed.

Ahermatypic corals. Differ from hermatypic corals in tropical areas in that they lack algae that live together with the corals and help build tropical reefs. Ahermatypic corals can live in cold and deep water.

Authigenic. Generated on the spot. As used in this book, refers to sediments consisting of minerals deposited directly out of solution by chemical processes.

Backwash. The seaward return of water following the uprush of a wave onto a beach.

Bar. A slightly submerged sand ridge.

Barrier. An emerged ridge of sand, ordinarily extending parallel to the coast.

Berm. The essentially horizontal upper portion of a beach.

Carbon-14. A radioactive isotope of carbon with atomic weight 14. It is used to determine the age of carbonaceous material not more than 30,000 years old.

Continuous reflection profiling. An acoustic method of making cross sections of the sea floor by using low frequency sound pulses that penetrate the bottom and reflect from various layers below the bottom.

Contours. Lines on a map, each of which connects points of the same elevation. The lines show the nature of the topography.

Cusp. One of a series of crescent-shaped mounds spaced at more or less even intervals along a beach.

Cyclothem. A series of beds formed during a sedimentary cycle and showing many repetitions of roughly the same sequences in the stratigraphic column.

Dendritic. Refers to a valley drainage pattern that resembles the branches of a tree or the lines of a broad leaf.

Fan-valley. A marine valley found crossing a fan-shaped deposit on the sea floor. Usually an outward continuation of a submarine canyon.

Fathom. Equal to 6-foot unit used for measuring depths.

Fault. A fracture in formations along which there has been displacement. Earthquakes are usually caused by a movement along a fault.

Fault scarp. A cliff produced by movement along a fault.

Glaciated. Formerly covered by glaciers that have now melted.

Groins. A short wall built out at right angles to the shore line to trap longshore drift.

Hanging valley. A tributary valley that enters the main valley at a higher level. If it were a land valley, there would be a waterfall at the junction.

Jetty. A relatively long wall generally extending out from shore to beyond the breakers. Used to protect harbors. Where two jetties are located on either side of the entrance to a harbor they serve to concentrate the tidal currents and keep the entrance deep.

Lamont. Refers to Lamont Geological Observatory of Columbia University, located at Palisades, New York.

Longshore current. A current of water moving parallel to the shore.

Moho. Short for Mohorovičić discontinuity, the juncture between the earth's crust and the underlying mantle where there is a considerable increase in the sound velocity of the underlying rock. Usually found about 5 miles beneath the ocean floor and 20 miles beneath the continents.

Rip current. A seaward moving current that returns the water brought in by wave action. These currents are likely to move at a higher velocity than a man can swim and hence are dangerous for swimmers. Rip currents move out along or near the surface, and should not be confused with the hypothetical *undertow*.

Scripps. Refers to Scripps Institution of Oceanography of the University of California located at La Jolla, California.

Transducer. A device for transferring electrical energy into sound energy. Used in echo sounding and in continuous reflection profiling.

Tsunami. A sea wave produced principally by sudden fault movements on the sea floor, which sweep across the ocean and at times build up to great heights along coasts, even thousands of miles distant from the source. The period between successive waves is usually 10 to 30 minutes.

Turbidity current. A current produced by sediment stirred up into a body of water, making the water heavier than that of the surrounding area. Hence, it moves down slopes and, according to some scientists, may attain high velocities.

Wave shadow. Where a sea wall, a jetty, or a point of land greatly decreases the intensity of the waves in the lee of the obstacle.

Weathering. Natural processes causing the decay and crumbling of surface rocks.

Woods Hole. Refers to Woods Hole Oceanographic Institution, of Woods Hole, Massachusetts.

GEOLOGICAL TIME SCALE

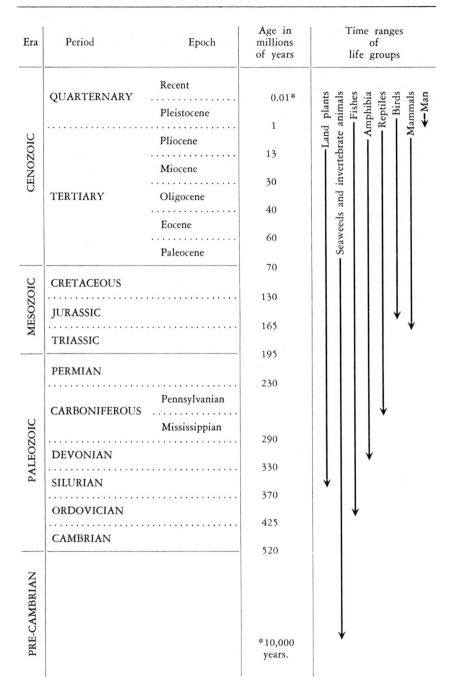

Era	Period	Epoch	Age in millions of years	Time ranges of life groups
CENOZOIC	QUARTERNARY	Recent	0.01*	
		Pleistocene	1	
	TERTIARY	Pliocene	13	
		Miocene	30	
		Oligocene	40	
		Eocene	60	
		Paleocene	70	
MESOZOIC	CRETACEOUS		130	
	JURASSIC		165	
	TRIASSIC		195	
PALEOZOIC	PERMIAN		230	
	CARBONIFEROUS	Pennsylvanian		
		Mississippian	290	
	DEVONIAN		330	
	SILURIAN		370	
	ORDOVICIAN		425	
	CAMBRIAN		520	
PRE-CAMBRIAN			*10,000 years.	

Life groups (columns): Land plants; Seaweeds and invertebrate animals; Fishes; Amphibia; Reptiles; Birds; Mammals; ←Man

SUGGESTED ADDITIONAL READING

(The following list includes books from which a wider coverage can be obtained on many of the subjects that are discussed in this book. Most of these books are of a more technical nature.)

Carson, Rachel L. *The Sea Around Us.* New York: Oxford University Press, 1951. 230 pp.
> A charmingly written, nontechnical book that was carefully compiled to make it as accurate scientifically as possible at the time.

Emery, K. O. *The Sea off Southern California.* New York: John Wiley & Sons, 1960. 366 pp.
> An informative book on all phases of oceanography in the interesting area off the southern California coast, with special emphasis on marine geology. Interesting but semi-technical.

Guilcher, Andre. *Coastal and Submarine Morphology.* New York: John Wiley & Sons, 1958. 274 pp.

Hill, M. N. (editor). *The Sea,* Vol. 3, New York: John Wiley & Sons, 1963. 963 pp.
> This volume covers the field of marine geology and geophysics of the sea floor. It contains articles by 41 authors, each writing about a separate field. Many of the articles are too technical for the non-scientist reader.

King, Cuchlaine, A. M. *An Introduction to Oceanography.* New York: McGraw-Hill Book Co., 1963. 337 pp.
> More up-to-date and less technical than *The Oceans,* and represents rather easy reading for most readers, having much less mathematics than most texts on oceanography.

Kuenen, P. H. *Marine Geology.* New York: John Wiley & Sons, Inc., 1950. 568 pp.
> Contains an excellent discussion of the ocean bottom and its history with abundant information from European sources.

Romanovsky, V., Bourcart, Jacques, and Francis-Boeuf, C. *La Mer.* Paris: Librairie Larousse, 1953. 503 pp.
> A splendid compilation of pictures and brief scientific descriptions from many sources. In French.

Russell, R. C. H., and Macmillan, D. H. *Waves and Tides.* New York: Philosophical Library, 1953. 348 pp.
> Scholarly but written in language understandable to the well-rounded layman.

Shepard, F. P. *Submarine Geology,* 2nd ed. New York: Harper & Row Publ., 1963. 557 pp.
> Gives much more complete discussion of all subjects presented in *The Earth Beneath the Sea* and also includes coastal classifications, mechanics of wave action and sediment transport (technical), methods of exploring the ocean floor, and synopsis of marine geochemistry.

————, and Dill, R. F. *Submarine Canyons and Other Sea Valleys.* Chicago: Rand McNally & Co., 1966. 381 pp.
> A complete description of all of the best-known canyons. Also includes summary of exploration by scuba and deep-diving vehicles. Complete discussion of theories of canyon origin.

Sverdrup, H. U., Fleming, Richard, and Johnson, M. W. *The Oceans.* New York: Prentice-Hall, Inc., 1942. 1087 pp.

Technical and somewhat difficult reading, but the most authoritative work on various phases of oceanography.

Trask, P. D. (editor). *Recent Marine Sediments.* Tulsa, Oklahoma: American Association of Petroleum Geologists, 1939, reprinted 1955. 736 pp.

Contains abundant information on various phases of marine sediments.

Whittard, W. F., and Bradshaw, R. (editors). *Submarine Geology and Geophysics.* London: Butterworths, 1965. 464 pp.

An international symposium discussing many recent ideas in marine geology.

INDEX

240

THE EARTH BENEATH THE SEA *(revised edition)*
by Francis P. Shepard

Designer:	Gerard A. Valerio
Typesetter:	Monotype Composition Company
Typeface:	Garamond and Futura Medium
Printer:	Universal Lithographers
Paper:	Glatco 660
Binder:	Maple Press